"This short, highly engaging book lets us into the backstage world of the workers essential to the live music industries. Through ethnography, interviews, and archival research, Kielich provides an intimate account of the labor which sustains the tours of major rock artists. With an eye towards the details of crew members' working days, their camaraderie and culture, gender norms, and the 'care work' of tour managers, *The Road Crew* places rock stars as bosses, crew members as part of the vast numbers of freelance gig workers animating popular culture. Kielich documents road crew work's informality and rules, its social costs and joys, its brutal schedules and surprising power relations. An essential book for anyone seeking to grasp the full scope of the labor of live music."

Shannon Garland, *University of California, Merced, USA*

THE ROAD CREW

The Road Crew: Live Music and Touring is an in-depth study of the road crew – the group of workers who handle the logistical and technical requirements of popular music concert tours – that provides an extensive look at the activities and personnel involved in the daily operation of these events.

Using interviews with road crew members, participant observation at concert venues and archival research, this book covers a range of topics, including how they learn their roles and maintain work through networks and informal practices, the experience of being on tour and the workplace culture of road crews, the daily tasks and necessary documents that contribute to the realisation of concert events, and the integral role that tour managers play in the working lives of musicians. The book also provides important insights into the experience of women working in a male-dominated field, the ways in which hierarchy shapes the working lives of "support" workers and the effects of touring on road crew members.

The Road Crew will be of interest to scholars and students of popular music, live music and the creative industries, as well as music fans, journalists, and professionals and practitioners in the music industries.

Gabrielle Kielich is a Senior Research Fellow at the University of Huddersfield. She has a PhD in Communication Studies from McGill University. In addition to touring and live music, her research interests include women and the electric guitar, rock music history and culture, the music industries and qualitative research methods. She has been a visiting researcher in the School of Culture & Creative Arts at the University of Glasgow and a course lecturer at McGill University.

THE ROAD CREW

Live Music and Touring

Gabrielle Kielich

LONDON AND NEW YORK

Designed cover image: Kkolosov via Getty Images

First published 2024
by Routledge
4 Park Square, Milton Park, Abingdon, Oxon OX14 4RN

and by Routledge
605 Third Avenue, New York, NY 10158

Routledge is an imprint of the Taylor & Francis Group, an informa business

British Library Cataloguing-in-Publication Data
A catalogue record for this book is available from the British Library

ISBN: 978-1-032-30017-7 (hbk)
ISBN: 978-1-032-30015-3 (pbk)
ISBN: 978-1-003-30304-6 (ebk)

DOI: 10.4324/9781003303046

Typeset in Optima
by codeMantra

For Dad

CONTENTS

FIGURES

ACKNOWLEDGEMENTS

This book began as my PhD thesis, and I would like to thank everyone who was involved. Since beginning the process of turning the thesis into a manuscript, I would like to thank Will Straw for encouragement and Rosemary Hill and Jan Herbst for support and helpful advice.

Many thanks to the editorial staff at Routledge who have been a pleasure to work with during this process: Hannah Rowe, Emily Tagg and Adam Woods.

This book would not have been possible without the people who participated. Huge thanks to all of the research respondents who took the time to share their knowledge and experiences in interviews: Andy, Craig, Tony, Duncan, Michelle, Adrian, Michael, Amy, Tasha and Ryan. Special thanks to Joe for being so generous and for all the inspiration.

A big shout out to the local crew at Venue A – thank you for the unforgettable opportunity.

I truly appreciate the support of my close friends. Big hugs and gratitude to Matt, Ann and James, Jess, and Eileen.

And, of course, I would like to thank my incredible family for continuous love, support and enthusiasm: my mom Mary, Grandma "J" June, Uncle Don and Aunt Renee', Uncle John, and Aunt Mary.

This book is dedicated to my father, Dr Lawrence Kielich.

INTRODUCTION

Road crews are essential and ubiquitous groups of workers in the live music industry yet have been minimally understood as subjects. The aim of this book is to demonstrate how the road crew is a key group of live music's support personnel. It will discuss their integral role in the realisation of concert events and provide an understanding of the variety of factors that shape their working lives and occupational identities. Prior to this study, the state of scholarship on road crews was correctly summarised by Sergio Pisfil (2020: 387), who observed "no serious work on road crews and their place in rock practices."[1] Such minimal attention was seen as unsatisfactory because of the important role road crews play in the economic, creative and experiential dimensions of live music (ibid.). By studying road crews, this book makes a contribution to existent live music studies[2] and provides an in-depth analysis of the "organizations and people that bring [concerts] into being," particularly those working backstage and on tour, and therefore creating a stronger focus on the "full range of actors, roles, and responsibilities involved in making events happen" (Anderton and Pisfil 2021: 4, 8). A history of live music without road crews is an incomplete account. The direct connection road crews have with concert events makes them an important source of insight into working in the industry. This study moves them to the forefront and, in doing so, widens the scope of who gets included in research and histories of popular music. More broadly, it contributes to wider scholarship that argues for the recognition and inclusion of a greater variety of support workers and their efforts as objects of scholarship in the cultural industries (e.g., Faulkner 1971, Negus [1992] 2011, Mayer 2011, Nakamura 2014, Maxwell 2016).

Looking closely at road crews brings into focus the importance of touring to the realisation of live music and its significance in their working lives. Touring is

DOI: 10.4324/9781003303046-1

a coordinated routine of transportation, accommodations and communication designed according to the strict confines of an itinerary but that is highly subject to change and revision. The experience of touring is an all-encompassing way of life that comes with expectations and effects. Touring shapes, constrains and informs the working lives, practices and relationships of a tour's personnel. Tour buses have particular rules; demanding schedules combined with close living quarters enforce certain types of behavioural conduct; and managing life back home requires adaptation and understanding. Sleeping schedules vary, as do the length of working days, and other people's priorities usually come first. The experience of touring may be mythologised as sex, drugs and rock and roll, but in practice is more directed to meeting basic needs than engaging in excess. This book draws attention to these features and works to develop and expand the understanding of touring and its relation to live music.

Musicians may fulfil the role of performers and the duty of performance in live music. However, the majority of people working in live music and on tour are not musicians, but rather the workers that assist them. That such people, and those working backstage, are often neglected is not unique to live music studies. As Norbert Elias (1978: 16) has described, people commonly speak of "structures as if they existed … above and beyond any actual people at all." Revealing backstage activity, notes a theatre scholar, also risks removing the "mystery and magic" of onstage performance (White 2014: 1, see also Webster 2015: 105–106). The concert space is a complex setting that is perceived simultaneously as routine and exceptional by differing parties (Frith 1978: 169; see also Webster 2011: 16). While the audience typically experiences a concert as a highly anticipated and exciting event, the road crew experiences it as the culmination of a daily routine that comprises their everyday efforts. However, road crews also understand and recognise the audience's perspective on the concert event and feel strongly connected to and motivated by its outcome. In attending to the routine elements of live music, they are very much a part of making it exceptional for the audience.

Acts of "innovation and creativity are not opposed to, but rather made possible by, the mundane cycles of the quotidian" (Felski 1999: 21). This book draws attention to the many kinds of activities that are involved in and important to the realisation of an event but may be separate from the creative and aesthetic decisions that directly shape live performance and the concert spectacle. It explores the taken-for-granted, quotidian and mundane – yet ultimately essential – aspects of concert events. The concert environment is so interesting because it blends a multiplicity of experiences in one setting and suggests there is ample room to expand and develop this area of study. The findings presented here provide a lens through which to understand the live music industry differently and demystify the extraordinary status accorded concerts and artists. The following section of the Introduction details the scope for the book.

Scope

Several factors need mentioning to outline the scope of the book. Live music and touring feature several types of crews: the local crew employed by the venue, the equipment suppliers' crew[3] and the artist's road crew. The emphasis here is on the latter. At times, the book discusses and refers to the venue's local crew in order to contextualise and explain particular aspects of the research (see also Kielich 2021). However, it does not make this group a central feature and does not engage in any sustained analysis of these workers. The book is focused on freelance, self-employed workers on road crews who are hired by and work directly for artists on tours.

An unexpected pattern, and a valuable finding, emerged during the research that informed its scope. All of the research respondents have (on average) 20 years or more of experience in their careers, and, as a consequence, the book is focused on an older and more experienced demographic of cultural workers.[4] The category of age brings with it the accumulation of experience and the ability for reflection. This aspect therefore presents a perspective from an understudied group of workers and a counterweight to the understanding of the cultural labour workforce as young (e.g., Hesmondhalgh 2013). It also problematises the claim made by Clem Gorman (1978: 18), in his behind-the-scenes account of rock concerts, that most crew members "do not stay on the road beyond the age of about thirty-five."

Time and place are also important factors in the range of the study. As implied by the above, this is contemporary research situated in the current time and recent past of the modern concert industry. It is not a historical account of workers on tour. The findings presented in the following chapters relate to working in live music from the mid-1980s to the present, and any occasional deviation from that time frame is noted. Regarding location, workers based in the United Kingdom and United States constitute the research sample. However, their working lives and experiences occur in an international setting, and mobility is a key component of their working lives.

The genres and types of tours covered are broad yet have particular emphases. Primarily featured here are road crews working for commercially successful and well-known artists in the rock music genre. The chapters also include findings from crews of artists in folk, jazz, R&B, hip-hop and electronica, as several workers have moved between them. Generally speaking, findings are related to mid-level tours, or in venues having an audience capacity of between 2,000 and 5,000. However, several factors complicate that narrow focus. Musicians may play a range of types of venues on the same tour and may alternate as support or headliners, which places road crews in diverse settings and broadens their work experiences. Additionally, over the course of their careers, road crew members may work for many artists on different types of tours, and, as such, my research findings draw on an accumulation of workers and experiences that likewise reflect this range, from small DIY tours in clubs to major production stadium tours. The book is less concerned with carefully describing a particular type of tour and is primarily interested in understanding the everyday experiences of workers on tour. The next section of the Introduction outlines important terminology as a way to establish the purpose of the book and the chapters that follow.

Key Terminology

It is important to understand the terms "live music" and "tour" as well as the relationship between them. As this book will demonstrate, the term live music does not only "refer to a commodity that is sold, bought, enjoyed, and accumulated ... but to the specific ways in which music is performed, staged, and mediated" (Anderton and Pisfil 2021: 4). Live music events are constructed through the coming together of various elements.[5] At the same time, live music is a *structure* that produces a tour, the process of touring and distributes important roles and duties. Live music can and does happen without touring: musicians play lengthy residencies in a single city, do "one-off" concert engagements or make a sole festival appearance. A tour, however, does not occur without live music. In this way, live music is the *purpose* of a tour.

From the 1960s, major record companies viewed tours as distinct marketing strategies and "promotional vehicles" for recordings (Laing and Shepherd 2003: 567, Negus [1992] 2011: 130, Frith et al. 2019: 2). The decline of record sales in the first decade of the twenty-first century (Hesmondhalgh 2013: 342) placed greater emphasis on touring as a major source of "bread and butter for most artists" (Black, Fox and Kochanowski 2007: 154, Shuker 2013: 49, Hracs 2015). However, most musicians "make their living selling their services as performers rather than from their returns from record sales or copyrights" (Frith et al. 2013: 62), and touring has been a consistent source of revenue independent of record sales (Laing and Shepherd 2003: 567). Country artists in the 1920s toured on the "major Southern vaudeville circuit," the working lives of jazz musicians in the 1930s and 1940s consisted primarily of "being on the road"[6] and the "chitlin' circuit" was an "economic necessity" for blues and R&B musicians that enabled African-American performers to "make a modest living" from the 1930s into the 1960s (Boyer 1990, Laing and Shepherd 2003: 567, Olsen 2007: 2, Peterson 2013: 45). In this book, shifts or distinctions in the function of touring are ultimately less important than what they can be reduced to: that a tour occurs when an artist needs or wants to perform live music as a means to support and/or develop their career (Black, Fox and Kochanowski 2007).

On a daily basis, on a tour, live music is the *objective*. The activities of the entire day are structured around the live music event and directed towards its outcome. This objective can be symbolised by a concert ticket[7] which grants an audience member access to an act of performance (Becker 1982: 119) that takes place in a particular setting at a specific time and that has been anticipated after being sold in advance (Straw 1999–2000: 155). Each performance is a singular event and an exclusive, immediate occasion for those in attendance. The concert ticket promises the possibility of a particular experience and, generally speaking, comes without the option for a refund should expectations fail to be fulfilled (Pitts 2014: 28, Brown and Knox 2017: 236).

A "tour" is the total set of all planned live music events in a particular itinerary (see Chapter 4). An individual concert "falls [within] a wider temporal series of events. The whole series is a called [a] tour" (Weinstein [1991] 2000: 203). Tours differ in scale, which will vary according to the types of venue and their capacity, along with production requirements and number of personnel. Tours of top-tier artists may consist exclusively of stadium or arena dates while mid-level musicians play large theatres, but they can also alternate between them. Tours may consist solely of concerts in urban venues but may also include festival appearances.[8] Furthermore, artists may spend part of a tour as a headliner and another part as a support act. These points are made to illustrate the variations that can exist when referring to a "tour," even as our understanding of a "tour" is rendered consistent by a series of events that comprise an itinerary.

The state of "being on tour" – or "touring" – refers to the experience and practices involved in realising a sequence of live music events. It is the process of bringing live performances to specific places at particular times. Touring is an activity and experience (something people do), a mindset (a way people think), a workplace (where people utilise skills, maintain a living and participate in its culture) and a social space (where people interact and contend with social phenomena). It is also an activity that is organised by and expressed through specific language and terminology. Touring creates, and unfolds within, a complex mobile world that exceeds any single event and is characterised by a particular way of life (Williams 1983).

The process of touring transforms places, localities and concert venues into temporary spaces in which important daily activities and preparations for performances are negotiated. For road crews, cities and venues are temporary and stationary hubs of activity that interrupt on-going mobility and are what Friedrich Kittler (1996) would call media that process and circulate workers from one place to the next. A significant part of touring occurs during that time and in movement from one place to the next, inviting an attention to how road crews work both *within* and *between* venues. A full understanding of live music means accounting for *mobility*. This statement builds on recent scholarship that has argued that "thinking about rock tours as *moving* productions brings out important aspects of rock live performance that still need further research" (Pisfil 2020: 389, original emphasis). Tim Cresswell (2006: 2) describes mobility as "displacement" and the "act of moving between locations" which is commonly exemplified by the idea of moving from point A to

B. Cresswell is particularly interested in the line that *links* A to B as a site of analysis (ibid., see also Nóvoa 2012: 366). His approach is to "explore the content of the line that links A to B, to unpack it, to make sure it is not taken for granted" (ibid.). This book follows Cresswell's approach to mobility in order to conceptualise and analyse this process and experience for road crews. Furthermore, it considers that the line between A and B shapes and is shaped by activities in both locations.

Live music from the perspective of the road crew means understanding how its members work alongside musicians or, rather, individuals with the occupational status and identity of stars, artists and performers. Such status and working relationships – and their differences – are a normal part of the everyday for road crews. Musicians' status as performers, artists and stars is integral to understanding the everyday life of live music. The use of these terms to identify and differentiate professional roles is not to invoke or reproduce stereotypical associations with these positions. Musicians encounter working conditions and face pressures associated with performing and touring (see Toynbee 2000, Jones 2012) that can and do influence everyday life and workplace interaction. Their absorption and "preoccupation with music and the life of the musician" is a further contributing factor (Jones 2012: 70). The presence of musicians occupying such roles is a component in how the hierarchy, conventions and daily activities of touring, as well as the experiences and occupational identities of road crews, are shaped.

Touring, as previously stated, is a significant aspect in the ways in which musicians earn a living, and live music forms the basis for this remuneration. In this way, musicians may be "particular sorts of workers" who are "people seeking to do jobs," but in the touring environment, their status as "performers, celebrities, and stars" (Williamson and Cloonan 2016: 8, 10) is central. While stars, for consumers, represent "a form of escapism from everyday life and the mundane" (Shuker 2013: 62), for workers in the live music industry, they are part of everyday life and the mundane. In addition to their roles as workers, performers and stars, musicians are also employers (Stahl 2013: 184), which is central to understanding their relationship with members of road crews. Accordingly, to bring road crews "into the glare of the academic spotlight" (Behr et al. 2016: 19) is to show how their role as *support personnel* is key to understanding them. The chapters will show that such secondary status influences the working lives of members of road crews, but that the notion of hierarchy on tour is not so simplistic. Rather, some crew members move between positions of authority and subordination that vary with context. Furthermore, this book calls into question Becker's (1982: 77) statement that the term support personnel, in its capacity to subordinate, "accurately reflects their importance in the conventional art world view." My findings challenge this notion by demonstrating their significance to the live music industry. The next section of the Introduction will briefly describe the book's approach.

Approach

This study sought to "discover and understand a phenomenon, process, the perspectives and worldviews of the people involved, or a combination of these" (Merriam 2002: 6). And to do so by exploring the "nature of that setting – what it means for participants to be in that setting, what their lives are like, what's going on for them, what their meanings are, what the world looks like in that particular setting" in the contemporary time (Patton 1985: 1 in ibid.: 5). My approach was to examine the working lives of road crews via a variety of methods and sources in order to optimise data and produce "knowledge on different levels" to achieve depth and understanding (Flick 2014: 184). The qualitative approach involved data collection through interviews with 12 members of road crews and workers in the live music industry. Half of the respondents agreed to be named and the other half chose to remain anonymous. In the interest of consistency, I used only first names here, and the first names of those respondents requesting anonymity have been changed. Fieldwork, or participant observation, took place during three months at two concert venues and included 30 concerts by 30 different artists. Venue A was the site of most of the fieldwork and is a 1,900-capacity venue and multipurpose facility, and Venue B is a 2,600-capacity music venue. Research also involved analysis of documents and archival materials, and consultation of primary and secondary sources and media representations. As such, the study involved triangulation, or the use of several methodologies or sources in the study of the same phenomenon in an attempt to enrich the study, capture a variety of dimensions and provide "a confluence of evidence that breeds credibility" (Denzin 1970: 291, Bowen 2009: 28, Flick 2014: 183–184, Eisner 2017: 110). Additional specificities of the various components of the research are detailed as relevant throughout the book chapters. The following section details my intention to the readership of this book.

A Note to Readers

This book is for, and will hopefully be of interest to, anyone who enjoys live music and is curious to learn more about the industry, its practices and its workers. It is for serious music fans and seasoned concertgoers and for those just starting to explore the world of live music. I hope this book will be of interest to industry practitioners, workers in live music and musicians, who may share or recognise the experiences recounted here, or who may disagree with aspects of the book and have divergent perspectives. This book began as my PhD thesis and therefore has its origins in and makes a contribution to the academic study of live and popular music. In this way, it is suitable for scholars and researchers of popular music studies, live music and media and communication studies, as well as graduate students and undergraduate students interested in these areas. However, in turning it into a book, it is also my intention that the content presented here is accessible to a much wider audience. I hope that anyone who chooses to read this book will take something away from it and into their next concert experience, and that it will inspire questions, conversations and new perspectives about live music, whether in the classroom, backstage or in the pub. The final section of the Introduction will describe the focus of each of the five chapters.

Chapter Descriptions

Each chapter focuses on an aspect of the roles, daily activities and experiences that comprise the working lives of road crews and shows how they are of importance to the realisation of concerts and to understanding live music. The book begins with Chapter 1, which defines and discusses who road crew members are and offers an analysis of some of the key characteristics of this group of support personnel. Chapter 2 continues this exploration by examining how members of road crews gain entry to the live music industry and manage careers as freelance workers on tours. These chapters provide important context for those who follow and make a contribution to wider scholarship on working and workers in the music and cultural industries.

The book then moves from defining road crews and important aspects of their career paths to exploring the complex mobile world of touring. Chapter 3 looks at the structure of a show day and the varied tasks and responsibilities occurring throughout. It also examines road crew members' temporary workspaces within concert venues and the significant activities occurring within them. Chapter 4 is about the experience of being on tour. It analyses the ways in which touring shapes the lives of road crew members both on and off the road, how they manage close working relationships and highlights important characteristics of their workplace culture. These factors offer deeper insights into the nature of touring and its effects on workers and, in doing so, draw attention to a wider set of practices and processes involved in live music events.

The book takes a close look at the specific role of the tour manager in Chapter 5. In particular, it examines the working relationship tour managers have with musicians. The tour manager's role is largely defined in relation to the needs and requirements of specific artists, and a key component of their jobs involves looking after musicians. The chapter considers what it means to do so and the ways that this working relationship shapes and potentially affects musicians as well as the decisions tour managers make, the nature of their daily activities and their own self-definition. In this way, this relationship is of particular importance to understanding the realisation of concerts. The Conclusion presents a synthesis of my research findings and offers suggestions for future studies.

Notes

1 See Battentier (2021) and Faraday (2021) for studies focused on the roles, significance and experiences of sound engineers in various contexts.
2 See the Introduction in Anderton and Pisfil (2021) for an overview of the field of Live Music Studies.
3 An equipment supplier is a company from which audio, lighting or video equipment can be rented for an event or a tour. The company hires crew members to attend to the rented equipment at the event or during the tour.
4 Richard Ames's (2019) book, a collection of personal interviews with some of the "pioneers" of the UK live music industry, presents additional evidence of an older demographic. He noted that, since the time they began working in the 1960s and early 1970s, over half of them "still work in the music business" while others had retired and some had died (x).
5 See Behr et al. (2016) for a discussion on an ecological approach to the construction of a live music event.
6 From 1935 to 1955, US jazz musicians faced restrictions from the musicians' union on touring in the UK. See Williamson and Cloonan (2016) for an in-depth study of the British Musicians' Union.
7 Tickets and attendant issues, such as pricing and the secondary market, are an important area of live music studies that is outside the scope of this book. See Krueger (2005), Brennan and Webster (2011), Budnick and Baron (2011), Johansson, Gripshover and Bell (2015), Behr and Cloonan (2020) and Westgate (2020).
8 Existent scholarship minimally examines touring circuits in terms of historical development and significance (Théberge 2005, Frith et al. 2013) and more contemporary considerations as to why musicians perform in particular markets (Johansson and Bell 2014).

References

Ames, Richard. 2019. *Live Music Production: Interviews with UK Pioneers*. New York: Routledge.
Anderton, Chris and Sergio Pisfil, eds. 2021. *Researching Live Music: Gigs, Tours, Concerts and Festivals*. London: Taylor & Francis/Routledge.
Battentier, Andy. 2021. *A Sociology of Sound Technicians: Making the Show Go On*. Wiesbaden: Springer VS.
Becker, Howard S. 1982. *Art Worlds*. Berkeley: University of California Press.
Behr, Adam and Martin Cloonan. 2020. "Going Spare? Concert Tickets, Touting and Cultural Value." *International Journal of Cultural Policy* 26(1): 95–108.
Behr, Adam, Matt Brennan, Martin Cloonan, Simon Frith and Emma Webster. 2016. "Live Concert Performance: An Ecological Approach." *Rock Music Studies* 3(1): 5–23.
Black, Grant C., Mark A. Fox and Paul Kochanowski. 2007. "Concert Tour Success in North America: An Examination of the Top 100 Tours from 1997 to 2005." *Popular Music and Society* 30(2): 149–172.
Bowen, Glenn A. 2009. "Document Analysis as a Qualitative Research Method." *Qualitative Research Journal* 9(2): 27.
Boyer, Richard O. 1990. "The Hot Bach." In *The Duke Ellington Reader*, ed. Mark Tucker, 214–252. Oxford: Oxford University Press.
Brennan, Matt and Emma Webster. 2011. "Why Concert Promoters Matter." *Scottish Music Review* 2(1): 1–25.

Brown, Steven Caldwell and Don Knox. 2017. "Why Go to Pop Concerts? The Motivations Behind Live Music Attendance." *Musicae Scientiae* 21(3): 233–249.
Budnick, Dean and Josh Baron. 2011. *Ticket Masters: The Rise of the Concert Industry and How the Public Got Scalped*. Toronto: ECW Press.
Cresswell, Tim. 2006. *On the Move: Mobility in the Modern Western World*. New York: Routledge.
Denzin, Norman K. 1970. *The Research Act: A Theoretical Introduction to Sociological Methods*. New York: Aldine.
Eisner, Elliot W. 2017. *The Enlightened Eye: Qualitative Inquiry and the Enhancement of Educational Practice*. New York: Teachers College Press.
Elias, Norbert. 1978. *What is Sociology?* London: Hutchinson & Co. Ltd.
Faraday, Jacob Danson. 2021. "Buried in the Mix: Touring Sound Technicians, Sonic Control, and Emotional Labour on Cirque du Soleil's Corteo." PhD diss., Memorial University of Newfoundland.
Faulkner, Robert R. 1971. *Hollywood Studio Musicians: Their Work and Careers in the Recording Industry*. Chicago, IL: Aldine.
Felski, Rita. 1999. "The Invention of Everyday Life." *New Formations* 39: 13–31.
Flick, Uwe. 2014. *An Introduction to Qualitative Research*. 5th ed. London: Sage.
Frith, Simon. 1978. *The Sociology of Rock*. London: Constable & Robinson.
Frith, Simon, Matt Brennan, Martin Cloonan and Emma Webster. 2013. *The History of Live Music in Britain, Volume I: 1950–1967*. Farnham: Ashgate.
Frith, Simon, Matt Brennan, Martin Cloonan and Emma Webster. 2019. *The History of Live Music in Britain, Volume II: 1968–1984*. London: Routledge.
Gorman, Clem. 1978. *Backstage Rock: Behind the Scene with the Bands*. London: Pan Books.
Hesmondhalgh, David. 2013. *The Cultural Industries*. 3rd ed. London: Sage.
Hracs, Brian J. 2015. "Working Harder and Working Smarter: The Survival Strategies of Contemporary Independent Musicians." In *The Production and Consumption of Music in the Digital Age*, eds. Brian J. Hracs, Michael Seman and Tarek E. Virani, 48–62. London: Routledge.
Johansson, Ola and Thomas L. Bell. 2014. "Touring Circuits and the Geography of Rock Music Performance." *Popular Music and Society* 37(3): 313–337.
Johansson, Ola, Margaret M. Gripshover and Thomas L. Bell. 2015. "Landscapes of Performances and Technological Change: Music Venues in Pittsburgh, Pennsylvania and Nashville, Tennessee." In *The Production and Consumption of Music in the Digital Age*, eds. Brian J. Hracs, Michael Seman and Tarek E. Virani, 115–130. London: Routledge.
Jones, Michael. 2012. *The Music Industries: From Conception to Consumption*. Basingstoke: Palgrave MacMillan.
Kielich, Gabrielle. 2021. "Fulfilling the Hospitality Rider: Working Practices and Issues in a Tour's Supply Chain." In *Researching Live Music: Gigs, Tours, Concerts and Festivals*, eds. Chris Anderton and Sergio Pisfil, 115–126. London: Taylor & Francis/Routledge.
Kittler, Friedrich. 1996. "The City is a Medium." *New Literary History* 27(4): 717–729.
Krueger, Alan B. 2005. "The Economics of Real Superstars: The Market for Rock Concerts in the Material World." *Journal of Labor Economics* 23(1): 1–30.

Laing, Dave and John Shepherd. 2003. "Tour." In *Continuum Encyclopedia of Popular Music of the World, Part 1: Media, Industry, Society*, eds. John Shepherd, David Horn, Dave Laing, Paul Oliver and Peter Wicke, 567–568. London: Bloomsbury Publishing.

Maxwell, Richard, ed. 2016. *The Routledge Companion to Labor and Media*. New York: Routledge.

Mayer, Vicki. 2011. *Below the Line: Producers and Production Studies in the New Television Economy*. Durham, NC: Duke University Press.

Merriam, Sharan B. and Associates. 2002. *Qualitative Research in Practice: Examples for Discussion and Analysis*. San Francisco, CA: Jossey-Bass.

Nakamura, Lisa. 2014. "Indigenous Circuits: Navajo Women and the Racialization of Early Electronic Manufacture." *American Quarterly* 66(4): 919–941.

Negus, Keith. (1992) 2011. *Producing Pop: Culture and Conflict in the Popular Music Industry*. Mountain View, CA: Creative Commons.

Nóvoa, André. 2012. "Musicians on the Move: Mobilities and Identities of a Band on the Road." *Mobilities* 7(3): 349–368.

Olsen, Allen O. 2007. "The Post-World War II 'Chitlin' Circuit' in San Antonio and the Long-Term Effects of Intercultural Congeniality." *Journal of Texas Music History* 7(1): 1–12.

Patton, Michael Quinn. 1985. "Quality in Qualitative Research: Methodological Principles and Recent Developments." Invited Address to Division J of the American Educational Research Association, Chicago.

Peterson, Richard A. 2013. *Creating Country Music: Fabricating Authenticity*. Chicago: University of Chicago Press.

Pisfil, Sergio. 2020. "Rock Live Performance." In *The Bloomsbury Handbook of Rock Music Research*, eds. Allan Moore and Paul Carr, 381–394. New York: Bloomsbury Academic.

Pitts, Stephanie. 2014. "Musical, Social and Moral Dilemmas: Investigating Audience Motivations to Attend Concerts. In *Coughing and Clapping: Investigating Audience Experience*, eds. Karen Burland and Stephanie Pitts, 21–34. Farnham, UK: Ashgate.

Shuker, Roy. 2013. *Understanding Popular Music Culture*. 4th ed. London: Routledge.

Stahl, Matt. 2013. *Unfree Masters: Recording Artists and the Politics of Work*. Durham, NC: Duke University Press.

Straw, Will. 1999–2000. "Music as Commodity and Material Culture." *Repercussions* (7–8): 147–172.

Théberge, Paul. 2005. "Everyday Fandom: Fan Clubs, Blogging, and the Quotidian Rhythms of the Internet." *Canadian Journal of Communication* 30(4). http://www.cjconline.ca/index.php/journal/article/view/1673/1810.

Toynbee, Jason. 2000. *Making Popular Music: Musicians, Creativity and Institutions*. London: Arnold.

Webster, Emma. 2011. "Promoting Live Music in the UK: A Behind-the-Scenes Ethnography." PhD diss., University of Glasgow.

Webster, Emma. 2015. "'Roll Up and Shine': A Case Study of Stereophonics at Glasgow's SECC Arena." In *The Arena Concert: Music, Media and Mass Entertainment*, eds. Benjamin Halligan, Kristy Fairclough, Robert Edgar and Nicola Spelman, 99–109. New York: Bloomsbury Academic.

Weinstein, Deena. (1991) 2000. *Heavy Metal: The Music and Its Culture*. Cambridge, MA: DaCapo Press.
Westgate, Christopher Joseph. 2020. "Popular Music Fans and the Value of Concert Tickets." *Popular Music and Society* 43(1): 57–77.
White, Timothy R. 2014. *Blue Collar Broadway: The Craft and Industry of American Theater*. Philadelphia: University of Pennsylvania Press.
Williams, Raymond. 1983. *Keywords: A Vocabulary of Culture and Society*. Revised ed. New York: Oxford University Press.
Williamson, John and Martin Cloonan. 2016. *Players' Work Time: A History of the British Musicians' Union, 1893–2013*. Manchester: Manchester University Press.

1

WHAT IS A ROAD CREW?

As the work activities of road crew members largely occur backstage and in the wings (but not exclusively), and their responsibilities involve working alongside stars and artists, they are subtly present yet always just slightly out of view. Members of road crews are "at work, right in front of the audience which largely ignores them in its lust for the stars" (Gorman 1978: 11). In this way, they are commonly referred to as "unsung heroes" who play important but rarely discussed roles in the lives of musicians and the organisation of live music. In the promotional campaign for Cameron Crowe's 2016 Showtime series titled *Roadies*, they were referred to as the "unsung heroes of rock." In the opening of his 2019 book, Richard Ames described the "pioneers" of the live music industry as "the unsung heroes of our business" (Ames 2019: ix Goldsmith). Road crews may be minimally understood, but they maintain a certain ubiquity in the cultures of rock and popular music. They are the subjects of several songs, are represented in films and television shows and there is even a beer named after them. Facebook pages are dedicated to specific crew members who look after well-known musicians, and artists commonly state their appreciation for crew members onstage and in interviews (see also Battentier 2021: vii). Furthermore, during the COVID-19 pandemic, road crews perhaps received more attention in the press and on social media than they ever had before. *Rolling Stone,*[1] *Billboard,*[2] *The New York Times*[3] and *The Guardian*[4] featured articles detailing the experiences of crew members during the pandemic and how to support them. It was during their absence from what they normally do that they became more fully recognised and appreciated as a result of their inability to do it.

This chapter will define, provide insights into and analyse some of the important characteristics of this group of live music workers. A road crew is not

DOI: 10.4324/9781003303046-2

simply the "people handling technical matters" (ibid.: ix) on a concert tour, nor a set of workers with undifferentiated skillsets. This chapter focuses on understanding the many and varied roles that comprise road crews and establishes their collective and individual nature. It will take a different view on the term "roadies" that has been commonly and colloquially used to identify crew members and their roles. I will discuss and problematise this term and show the ways in which the acceptance of the word has changed amongst crews due, in part, to particular associations that are largely outdated. The chapter presents findings that will situate contemporary meanings and usages of the term within their workplace culture.

Road crews come into existence for the purpose of producing a concert tour and realising a series of live music events. In doing so, the activities and occupational identities of members of road crews become entwined in the hierarchical structure of the live music industry. As crews form around particular artists and their specific career endeavours and touring needs, the members of road crews become positioned as "support personnel." Being support personnel is a key aspect of understanding the working lives of crew members and, as this chapter will show, is a complex and nuanced term within the touring workplace. Finally, a significant feature of road crews is the gender imbalance and minimal diversity that constitutes them. Live music, like so many sectors of the music industries, is predominantly a white, male-dominated field, and this factor has effects that permeate the workplace culture and working lives of road crew members. I will introduce this issue in this chapter, though the topic of gender will be a consistent theme throughout the book. Overall, this chapter specifically establishes key aspects of road crews and their significance to live music while also contributing a deeper understanding of support personnel working in the music and cultural industries more broadly.

What Is a Road Crew?

A road crew comprises individual workers with specialisations who handle the logistics and technical requirements of a concert tour. They collectively work for the realisation of live music events and do so in ephemeral and temporally bound formations. The composition of a road crew can be understood by the features of crews more generally. A "crew" is defined as a "group of expert specialists each of whom have specific role positions, perform brief events that are closely synchronised with each other, and repeat these events across different environmental conditions" (Sundstrom et al. 1990, Klimoski and Jones 1995, Webber and Klimoski 2004: 265). A crew forms temporarily to carry out a specific task, and after an assignment is complete, the members disperse and take new work in units comprising different members (Arrow and McGrath 1995: 380, Webber and Klimoski 2004: 266). Interviews with research respondents confirmed this:

> ... the industry is staffed by very highly skilled, highly experienced professionals who do a job that only a very small number of people in the world could actually do at the level that I was doing it at ... when I quit ... probably ... there was 100 people in the world that had the experience to do the job that I was doing.
>
> *(Craig 2017)*

> What people don't tend to realise about people who are in this job is that they tend to be exceptionally technically knowledgeable who live an incredibly transient life and are able to socially survive in this environment.
>
> *(Tony 2018)*

Members of road crews work, live and travel together throughout a tour's itinerary and duration, and participate in shared norms and a workplace culture. As they comprise distinct roles that function interdependently, working on a road crew means occupying an individual position with unique responsibilities at the same time that it involves participation, cooperation and identification with a larger group. In this way, understanding a road crew involves balancing the view that they are "competent individuals coming together to do work" with a perspective that accounts for and acknowledges the characteristics of the group (Ginnett 2019: 80). The size and specialisations of a road crew vary according to the production specifications and budget of a tour and are a function of the personal and professional requirements of musicians. Figure 1.1 lists common roles on a road crew.

A crew forms when "an organization has developed the technology to carry out a class of projects (tasks) and then selects personnel to staff the project" (Webber and Klimoski 2004: 266). Following this, road crews come into existence when an artist is ready to go on tour, requires specific personnel to assist them and has the resources to afford them. In the early stages of a career, often originating at the local level (see Bennett [1980] 2017: Chapter 4), musicians deem hiring someone a worthwhile forfeiture of part of their pay due to how it can save them time and energy on preparations and transportation. H. Stith Bennett ([1980] 2017: 74–76) notes that, contrary to the misnomer that musicians' working lives are "easy" and "devoid of manual labour," the need to move and attend to equipment means that hiring someone is a "common solution to the frustrations of physical work." These rudimentary requirements can be extended and applied to artists who attain higher levels of success and greater resources. When it becomes possible to develop more complex live shows, the need develops for a group of more highly skilled and specialised workers. Hiring road crews means that the labour associated with live music is distributed across several parties who collectively share the responsibilities and work towards its realisation. The next section of this chapter will offer an analysis of the term "roadies" that is used to colloquially refer to road crew members.

Tour Manager (TM): TMs have many responsibilities that vary according to the needs and requirements of artists and tours. Chapter 5 will explore their working lives in detail. Key components of their job are to handle the tour's budget, advance[1] each concert date, and maintain the smooth operation of a tour by attending to day-to-day activities and troubleshooting problems as they occur. This largely involves communicating information about schedules, transportation and accommodation to the touring party and doing so in a manner that ensures prompt arrival at all relevant destinations. Depending on the size of the tour, some of these tasks may be divided between other roles, including the production manager or production coordinator. An important aspect of the TM's role involves working closely with musicians. TMs travel with musicians, oversee their promotional tasks, sometimes handle their personal affairs, and maintain communication with artist management. They also pay per diems to the artist and crew, settle the artist's performance fees with promoters on show days, and report attendance numbers to the artist's management (Reynolds 2020). TMs also rely on and communicate with local crews at venues on show days.

Production Manager (PM): PMs oversee technical workers on the road crew and handle all aspects of a concert's production requirements, including sound, lights and video equipment, as well as their transportation and budget. PMs handle the technical riders that detail specifications and are concerned with the logistics of load-in and load-out at venues on show days. Depending on the size of the tour, the PM may also handle the travel and meal requirements for the road crew. PMs do not need to be technical experts themselves; rather, they oversee the activities of skilled technical workers. Communication between the PM and the TM is essential.

Road Manager: This term used to be interchangeable with "tour manager" but is now more closely associated with the handling of personal needs for musicians and attending to hospitality. Some touring parties will hire a road manager (in addition to a TM) who will manage the responsibilities associated with hospitality and the personal requirements of musicians.

Production Coordinator: On larger tours, a production coordinator will absorb some of the responsibilities of the PM. Their duties vary according to the tour and the working preferences of individual PMs and TMs. Coordinators can be responsible for handling non-technical aspects for the road crew, such as booking travel, dealing with immigration, ensuring receipt of per diems, and organizing food needs and catering.

Head of Security: Security is present on tours of very high-profile artists or when it is deemed necessary for the musicians' safety. They have a range of responsibilities that vary according to artist, including close protection services and directing safety measures. They are involved in ensuring the artist's safe arrival and departure to and from venues, hotels and transportation. In addition, they coordinate with local authorities, handle security logistics before, during and after performances and issue accreditation/passes.

Front of House Engineer (FOH engineer): The FOH engineer is responsible for mixing and controlling the audio that the audience hears during a concert, and ensuring good sound quality. The FOH engineer also handles all of the audio equipment and oversees the technical crew. During a concert, the FOH engineer is located behind a console in a designated area in the middle of the audience, and faces the front of the stage.

FIGURE 1.1 Common roles on a road crew.

Instrument Technicians/The Backline (e.g., Guitar, Bass, Drums, Keyboard): Instrument technicians are responsible for the daily maintenance, operation, set up and packing up of musical instruments and their associated audio components. They also oversee the musicians' stage environment, such as the set list, towels and water. Instrument techs attend to any issues that may arise before a concert and are positioned in the wings during a performance in order to troubleshoot as needed.

Lighting Designer: The lighting designer conceptualises the lighting display for the concert, creates a plan that specifies the types and order of lights, and works in conjunction with technicians to arrange the program controls of the lighting system. On tour, the lighting designer can run the show or assign the task to someone else (see Reynolds 2012).

Monitor Engineer: The monitor engineer mixes the sound that musicians hear while performing. This involves creating a tailor-made mix according to each musician's preferences through monitor speakers or in-ear monitors. During a concert, the monitor engineer is normally positioned at the side of the stage in order to communicate with musicians, which is typically done with hand signals.

Rigger: Riggers operate the rigging system and ensure the safety of the equipment and people in attendance. They assist with hanging and securing production elements of a concert by determining where lifting motors or cables should be attached. These are then used to lift the production requirements, such as lighting, sound and video equipment, above the stage accordingly.

Caterer: The caterer is responsible for coordinating and preparing meals for the touring party during a show day. Some tours carry their own catering; in other cases, catering is provided at the venue by local staff in accordance with details provided in the hospitality rider and in communication with relevant touring crew.

Wardrobe: Wardrobe personnel handle the stage clothing requirements for musicians. They attend to the care and maintenance of clothing on tour and assist musicians with any clothing or style-related preparations ahead of their performances.

Merchandise: Merchandise staff, known colloquially as "merch," are responsible for handling the stock and selling an artist's merchandise, such as t-shirts and albums, during concerts. They maintain an inventory, pack and unpack all merchandise items before and after shows, work the merchandise table or booth during concerts, process payments from customers, and handle the accounting of sales each night.

Bus Driver: The bus driver is the person who drives the tour bus of the musicians and/or crew during a tour. They are responsible for driving and maintaining the bus and working according to set departure and arrival times at venues, hotels or other locations. The bus driver's schedule is dictated by the tour's itinerary, the structure of the show day and the artist's schedule.

[1] Advancing refers to the process of communicating with venue promoters and relevant personnel to discuss and confirm show details and logistics and ensure that technical and hospitality requirements will be handled correctly. It also involves obtaining relevant information about the venue's location and spatial characteristics, and establishing times for arrival, soundcheck and performance (see Reynolds 2020). For the TM, advancing is a way to prevent problems that may occur on a show day. By gaining awareness of what to expect, they are able to anticipate potential problems or advise the touring party on how to attend to them, prior to the event (ibid.).

FIGURE 1.1 (continued)

"Roadies"

Members of road crews are commonly referred to as "roadies." The term is familiar enough in popular culture that 12 December 2012 was declared "international roadie day" given the 1-2-1-2-1-2 numerical configuration of the date and its reference to the process of sound checking (Virtue 2012). It is also reflected in the titles of topical films,[5] television series[6] and songs about them[7] (see Battentier 2021: vii–ix). The term is generic and can be used to describe any member of a road crew. Deena Weinstein (2003: 565) articulated the meaning of the term.

> As rock concerts have become more sonically and visually elaborate, the number and specialization of roadies have increased accordingly. Roadies set up the show, maintain (repair and tune) the musical instruments, operate equipment during the show (the lights, sound and special effects), and pack up ('tear down') and transport everything required for each night's concert. Bands playing at very small venues serve as their own roadies or take good friends on their low-budget everything-in-a-van tours. In contrast, arena-sized rock tours employ dozens of roadies, many of whom are well-trained and sometimes well-paid specialists. Each musical instrument has its own 'tech,' and those in charge of sound and lighting ('engineers') increasingly have had formal training. Other roadies are selected solely for their brawn, working as 'grunts' or security staff.

Based on this passage, and in view of the definition of a crew as comprising specialised roles, to identify an individual member of a road crew as a roadie is a contradiction in terms. If a crew is defined as such, then central to understanding road crews is accounting for and making sense of the individual roles that comprise them. Weinstein's passage is useful for how it clearly illuminates problems with the term roadie. On the one hand, the difficulty is essentially based on a question of language. In referring to "techs" and "engineers," Weinstein acknowledges the variety and specialisations of roles on concert tours. However, on the other hand, in foregrounding the term roadie in characterising these workers, she denies the importance of specialisation and undermines the specificity of their roles. The definition of roadie is further confused in relation to skillset and differing scales of tours. If roadies can be "grunts" hired solely for their brawn, and the role can be fulfilled by the friends of early-career musicians,[8] it is hard to grasp how the same term can be effectively applied to the "specialists" that have become integral to advanced concert technologies on arena tours. A roadie is associated with the *general* responsibility and experience of working for musicians on tour while the term also invokes *particular* roles. The universal application of the term renders it meaningless in understanding what these workers actually do and are responsible for.

"Roadie" was once the operative term for musicians' support personnel on tours (Bennett [1980] 2017: 75). Any member of a touring party could be referred to as a roadie. For example, Peter Hince (2011) was called Queen bass player John Deacon's "roadie," The Baker was called The Clash's "drum roadie" though he handled multiple other roles and Tappy Wright (2009), a TM/road manager for The Animals and others, often self-referenced as a "roadie." TMs at times were distinguished from roadies, as in this description of a 1973 tour's entourage: "the whole group, crew, band and roadies everything, everybody, tour manager, probably 22 people" (Ames 2019: 21). At the same time, TMs could also be called "head roadies" (Hart 2011: 37–38). In this way, "roadie" presented limitations for crew members' occupational identities as well as in their capacity as objects of study. The problematic nature of the term roadie is reflected in and supported by statements from road crew members. Tana Douglas (2021: 100) confirmed this during the early part of her career in the 1970s.

> the public had no idea what we did apart from hump boxes and … Aside from 'roadie,' the only other official title was 'road manager' … More titles were emerging as the size of productions grew, but in those days we did everything … although each of us specialised in different parts of the job, we were all just roadies.

Research respondents indicate that as concerts have become more elaborate and crew members more specialised comes the desire to be recognised and perceived accordingly. They acknowledge that language is a key component in such occupational identification, and while their own perception of and commitment to their work has not changed, the importance of the terminology used to describe it has.

> … it's funny cause we used to be called roadies, but now everybody wants to be a technician don't they? … everybody wants to be a technician and taken seriously whereas I was still quite serious when I was being called a roadie. I think the thing is if you use the word roadie people don't see it quite a proper thing as being a technician do they.
>
> *(Duncan 2018)*

> I don't use it myself. I call myself a technician. I will use the word roadie kind of like in jest, but I wouldn't use it in any professional way.
>
> *(Adrian 2018)*

The generic nature of the term and, likewise, the wide range of skills associated with roadies are incompatible with the working practices of research respondents in this study. Simon Frith et al. (2021: 48) provide additional evidence with a quote from the Stereophonics' stage manager, who confirmed that

> People aren't just roadies anymore; there's no such thing as a roadie, you can't do the job now if you're just a roadie. A lot of people are very highly qualified mathematicians or physicists, or really highly qualified electrical engineers, who transgress into this business now because they like the lifestyle, they get paid very well and we need them! [laughs] … If you think you can just hump a box to get a job, forget it. That's not the qualities we're looking for anymore …

The physical aspects associated with the responsibilities of a roadie essentially refer to "humpers," which is jargon for people who haul equipment and push flight cases during load-in and load-out. This type of unskilled labour would not feature in a larger division of labour on contemporary tours. Humpers do exist, but it is the local crew, not the touring crew that fulfil this duty.

> There are no unskilled people on the road cause we can't afford to take unskilled people on the road … Local crew will cover the load-in, load-out, the humping if you like. They'll load and unload the trucks, they'll help with unpacking things out of the flight cases, they'll help set things up, the risers, things moving about on stage, they'll put all the flight cases away somewhere neat, they'll usually know in advance where they go that doesn't block fire exits.
>
> *(Joe 2018)*

This activity was evident during participant observation at Venue A and Venue B and was attended to by the local crews. The distinction can be further observed in Roy Shuker's (2008: 58) overview of the concert environment, in which he describes the range of personnel working backstage. He states that there are "technicians in charge of the instruments and equipment (amplifiers, etc.); stagehands, *who often double as roadies*, people to work the sound and lighting boards, security guards, and the concert tour manager" (ibid., my emphasis). As such, the term roadie is laden with meanings that deflect or undermine the skillsets of road crew members.[9]

Research respondents report that they now perceive the term as "pejorative" (see also Douglas 2021: 298), use it solely in humour and do not believe the term accurately reflects what they do.

> Cause that's what people think, 'oh what do you do?' I work in the concert industry, I'm a technician in the concert industry. 'Oh is that like a roadie?' … I don't find it a helpful term. It doesn't tell you anything about what a person actually is or what they do.
>
> *(Adrian 2018)*

Only two respondents self-referenced as such in interviews. During participant observation, only one member of a road crew used the word roadie, and the local PM used the term once, in specific relation to a particular touring party. Taken together, research findings suggest that the term is now rarely used by and less accepted amongst members of road crews as a label for their occupational identities. In this way, the work itself has not changed as much as the language used to describe the work has taken greater importance.

Not all members of road crews, however, reject the term "roadie" as an occupational identifier, nor do they agree with the notion that it is belittling. The website and Instagram account "roadiedictionary"[10] provides definitions of workplace jargon. Its definition of roadie is as follows: "Did you know there are a bunch of people that get offended by this word? They find it derogatory?" (Thomas 2020). That such a description is favoured over an actual explanation of the term suggests that the rejection of "roadie" is not only widespread but also contested. It could also be a reference to a shift in the term's meaning and usage. Matt McGinn (2010: 39), longtime guitar technician for Coldplay, stated in his book that some "roadies refuse to admit they're even 'roadies' at all" and prefer, consistent with research respondents in this study, to be referred to as "techs" or "technicians." McGinn understands this viewpoint due to the former term's association with a period of time during which tours were characterised by excess, and "roadie" can therefore imply a lack of professionalism (ibid.). The term also "dated people in a time where everyone needed to feel current and had to have an official job title" (Douglas 2021: 298). Adrian (2018) further confirmed how the term roadie is used and what it implies on contemporary tours by stating that "when you say that word it does immediately make those associations in people's heads."

"Those associations" is a reference to a particular, and largely antiquated, lifestyle associated with touring and being a "roadie." This shift in terminology is a window of insight into a broader change within the culture of work on tour. It could be summarised simply by the cliché of "sex, drugs and rock and roll" that permeates understandings of rock music history and has been expressed and mythologised in books, films and documentaries[11] that recount rock tours and in autobiographies written by road crew members. A feature in *The Guardian* succinctly summarised this era.

> Before the live-music industry became a billion-dollar behemoth, being on the road was, for many bands, a wild west of sex, drugs and even some rock'n'roll. Hedonism was rife, and it wasn't just the musicians who pillaged. Their road crews were right there with them, benefiting from a macho atmosphere where the expectation was that after they had unloaded the gear they would match their employers in debauchery … Roadie annals are full of such stories, many of them involving unpleasant treatment of female fans.
>
> *(Sullivan 2014)*

Much of this excess and hedonism was associated with the 1970s era of affluence within and generous tour support from the record industry (see Eliot 1989). Roadies were associated with bad behaviour which was tied to the culture of touring and the more general stereotypes and excesses of rock music culture. Touring at the time was seen as "all about the fun and games" and crew members "worked hard and you played hard and you tried to learn as much as you possibly could along the way" (Douglas 2021: 116–117).

This was during the time when the practices of the modern live music industry were still being established and pre-dated the decline of record sales that created new emphasis, and pressures, on live music as remuneration. Evidence of a culture change was observable during fieldwork when particular crew members who had worked across several decades referred to earlier eras as "the good old days" and made reluctant comments such as "you actually have to work now" which identified a perceived change in norms, practices and expectations for crew members. In particular, these crew members were lamenting a shift in attitudes about and tolerance of alcohol consumption on the road. Such changes were further confirmed and commented upon by research respondents and additional sources.

> I would say that boozing while working is phasing out much to my delight. You still get the few people that do it, but it's now becoming less common, it used to be very common, people would just quite happily have a few in the morning and a few in the afternoon and a few in the evening and that would just be part of it, certainly the same with drugs while working … Yeah, the crew would do it. We don't so much now. You still get some old school people who do, it's just a different way of doing things. Most crews now would fire somebody for doing that. Some crews tolerate it, some crews encourage it.
>
> *(Joe 2018)*

> Bad behaviour isn't acceptable anymore, to be drunk and carrying on … A lot more is expected of you. People think it's crazy backstage, and it's girls and drugs, but it's not. It's people working and having a cup of tea.
>
> *(Chris McDonnell, Sound Engineer in Sullivan 2014)*

These changes in culture and attitudes about appropriate conduct are the result of a number of factors: the reduction of tour support by record companies (see Shuker 2013: 59), the technologisation of tours (see Cunningham 1999), stronger emphasis on health and safety regulations and live music becoming a central site of income generation.

Douglas (2021: 100) noted that roadies "copped a lot of flak back in those days." During this time, there was a "stigma attached to our position" – though some of them "revelled in it." Roadies were perceived as "a bunch of wild, tough, and not necessarily talented, yobs" in the days "before you could go to school to get a degree on how to be a roadie. (Oops! Sorry – technician)" (135). She notes, however, that such an impression was "inaccurate" and roadies were "innovators who knew no fear or boundaries for change while creating the new face of the music industry" (ibid.). However, if you "came from the era of 'roadie,' you weren't in step with new technology" despite having been associated with creating its path (298). Following this, the term is no longer consistent with much of the realities of working on a crew and retains an outdated association. The term "roadie" refers to a particular time and a specific understanding of what a member of a road crew did. By rejecting such a term in favour of more specialised occupational identifiers, crew members distance themselves from these past associations. Furthermore, the specialised roles reflect the advancement of the live music industry.

It is worth noting that an unflattering association with "roadies" can sometimes still be observed that can undermine the significance of their roles. For example, the perception and understanding of road crew members is not helped by instances such as the volunteer scheme at the Rock and Roll Hall of Fame and Museum in Cleveland, Ohio, which used to call its volunteers "Roadies." This effectively equated the skill and value of a group of workers – who support the industry and artists the Hall of Fame seeks to celebrate and preserve – with unpaid labour and reproduced expectations of compensation for cultural workers more generally (see Banks 2007, Hesmondhalgh and Baker 2011, Umney and Krestos 2015). Additionally, informal observations reveal the regular occurrence of audience members at concerts mocking and shouting expressions of impatience towards road crew members working onstage in preparation for the show. Such activity seems to suggest a more generalised lack of awareness about their roles and the importance of their responsibilities.

Despite all of this, McGinn (2010: 41) is "proud, not ashamed, to call myself a roadie and always do," and he notes this is "much to the amusement/horror" of those who distance themselves from the term. Significantly, he believes that crew members want to reclaim "roadie" as a "positive term and revive it, with good association" (42). Evidence suggests that this is true and is occurring. Though crew members reject "roadie" to describe their individual roles, they continue to utilise the term as shorthand to refer to the collective experience of being a worker on touring crews. The word "roadie" is used to signify that a person or group is part of this specific line of work.

This usage of the term can be observed in various types of communication. Books written by road crew members often include "roadie" in the titles. For example, *Roadie: My Life On the Road with Coldplay* (McGinn 2010), *Rock Roadie* (Wright and Weinberg 2009) and *Roadie, Inc.: How to Gain and Keep a Career in the Live Music Business* (Reynolds 2012). It is further evident in media forms that broadcast their stories and causes, such as the podcast *Roadie Free Radio*[12] that features interviews with crew members and sells roadie-themed merchandise, and the album project *Whole Lotta Roadies*[13] that was created to raise funds during the COVID-19 pandemic. There is also the Australian organisation Roady for Roadies[14] that likewise raised awareness and money for crew members affected by the pandemic. The Roadie Clinic[15] is a US-based non-profit that offers services and resources to road crew members who may struggle with the difficulties of life on the road. Social media accounts also reflect how the term "roadie" still circulates, such as the "theroadiecookbook,"[16] which offers a space for crew members to share stories and recipes, discuss food they have tried on the road and raise funds. Workplace terminologies such as "Roadie Friday," which refers to a show day before a day off on tour, further substantiate its continued use.

These examples suggest that crew members have reappropriated the term "roadie" and reserve its use for a means to express a more general participation in the work experience of touring rather than as a label for their specific role within the collective road crew. The next section will explore the idea of "support personnel," what it means to occupy this identity for members of road crews and how it functions for workers on tour.

Being Support Personnel

Touring is organised around the musicians and live music performances that are at its centre. Artists perform the "core activity" and make the "choices that give the work its artistic importance and integrity" while support personnel are an essential group that assists them and contributes to the outcome by handling all of the other necessary tasks (Becker 1982: 24–25, 77). The term "support personnel" implies a subordinate position or secondary status. To be or give support at once invokes the nature of work activities and the question of hierarchy.

Support is indeed assistance, but it is also productive and constitutive: an active process of "holding up," "enabling" and "looking after" (OED 2020). The position taken here is that road crews are *primary workers in supporting roles*. While they are support personnel to artists, they are also primary workers due to the significance of their supporting efforts in realising live music. The use of the term "primary" is adapted from David Hesmondhalgh's (2013: 259) characterisation of creative personnel. The latter are "primary workers in the production of symbols, information, entertainment and meaning – the core products of the cultural industries." If live music is the structure, purpose and objective of a tour, and it distributes important – "core" – roles and duties that must be fulfilled and enacted by a group of specialised workers in order for it to be realised, then road crews are primary workers.

If "the number of personnel reflects the size of the tour and the economic importance of the musicians" (Shuker 2013: 49), then the road crew's presence and work on tour are symbolic of another group's success. Seen another way, the road crew enable and contribute to musicians' success, and its continuity, through their supporting roles.

> There's also a satisfaction in helping a band progress. It's not always the case, sometimes you join a band and they're already at a high level and you just keep 'em going, but the last band that I toured with was [artist], when I started working with them they were quite raw and not that great live and needed a lot of technical help to step up to the level that they were trying to get to, and seeing them go from the initial gigs that we did with them which were like, I think the first gig I did with them was at a 250-capacity club in London to two nights in the Hollywood Bowl, you know 20,000 people and you know being and totally owning it. Being part of that and I think if you're a monitor engineer you know the band are there every night, you're right there with them every night, so it's quite an intimate relationship and so being part of that and going through that journey it was incredibly satisfying and I did that with a few bands. So that was great.
>
> *(Craig 2017)*

Contributing to the development and advancement of an artist's career in the live setting is an integral component of the working lives of road crew members, and the ability to witness and experience such a process fosters a sense of gratification. These efforts also illuminate the ways in which the work and responsibilities of road crew members function as a form of care. The topic of care, and how it operates on tour, will be discussed in greater detail and demonstrated most clearly in Chapter 5 in relation to tour managers and their responsibilities. However, it is worth establishing here that the roles of road crew members can be seen through a caring lens. At its most essential, care is a concern for, or action directed towards, others. It suggests a "reaching out to something other than the self" and positions another person's needs as the "starting point for what must be done" (Tronto 1993: 102, 105). This notion corresponds with and complements the roles of support personnel as workers tasked with assisting and enabling the activities of musicians.

Findings presented in this book indicate that there are significant intersections between notions of care and the occupational identities of support personnel. I would like to suggest that the idea of care advances the understanding of support personnel, their definition and what they do. Research presented here substantiates how care and support personnel have been theorised as elements that destabilise the notion of creative work as individual (Becker 1982, Alacovska and Bissonnette 2019), which foregrounds the collective and interdependent nature of live music. As findings in this book demonstrate, the activity of "care" and occupational identity of "support personnel" are both marked by the potential to be subordinated and relegated to secondary status and to be actively responsible for enabling and taking care of others. Being a support worker and doing care work involve putting the needs of others first, which is determined by and reproduces status and privilege. And finally, research in this book highlights that care also relates to questions of health and wellbeing, both in terms of self-care and care for others, along with questions of responsibility for others. In these ways, support personnel have much to reveal about the networks of people and activities necessary to produce cultural products.

On tour, crew members attend to an array of necessary tasks that are integral to the realisation of live music and that contribute to musicians' ability to successfully perform. They help musicians navigate an environment that is changing, largely unfamiliar and potentially intimidating (Leyshon 2009 in Battentier 2021). The presence and responsibilities of crew members are designed to facilitate musicians' ability to focus on performance, create a stress-free and consistent sense of everyday life on tour and handle the required logistical and technical tasks that are fundamental for live music to occur. These factors position the efforts and work activities of crew members as a form of care.

In their capacity as support personnel, road crew members can be perceived as "invisible" workers due to the ways in which their tasks and activities generally occur behind the scenes. As with other types of support workers in the cultural industries, the labour of road crews makes an important contribution but often goes unnoticed, unrecognised or unacknowledged (see Mayer, Banks and Caldwell 2009). In the film industry, this would be associated with the notion that "everything that matters happens on-screen, not off" with off-screen workers and their labour being "largely invisible to the general audience" (Curtin and Sanson 2017: 1–2). Within the live music setting, the responsibilities of road crews mean they are directly involved in the realisation of concerts, yet are ultimately excluded from the spectacle itself. While road crews are "invisible" in the theoretical sense of the term utilised by cultural industries scholars, as should already be clear, I do not support the view that activities that happen out of sight do not matter but are rather constitutive of live music events. Furthermore, this book demonstrates that a group of workers' relative visibility to audiences should not be a determining factor in whether or not they are perceived as important.

A more precise way of describing road crew members is that they are *less visible* given their occupational expectations and the functional components of the realisation of live music events. Road crews are ultimately workers who are not supposed to be seen, or should be minimally visible, in order to avoid detracting from musicians' performances. This is enhanced by the clothing that members of road crews wear, which generally consists of a black shirt (sometimes provided by the artist), black shorts or trousers and black boots, but does vary. Road crew members' attire has often been the target of humour and stereotypes because they have been known to wear old band t-shirts and worn-out clothing. Their clothing choices have been referred to as the "Roadie Uniform" and have been the subject of mythologising and mockery in the press (see Simpson 2009). However, black clothing is an observable, consistent and functional characteristic of stagehands and behind-the-scenes workers across cultural sectors. The manner in which members of road crews choose to express their version of this convention is connected to participation in a music culture, yet style choice ultimately retains a utilitarian purpose. Black clothing contributes to making crew members less conspicuous and is a practical choice given that "there's a certain baseline of what you're wearing is going to get dirty and torn" (Joe 2018), and stronger shoes are safer in a concert environment. The minimal visibility of road crew members is also equated with a job well done.

> My job role, you should take pride in position if nobody knows who you are, if you're never in a photograph or very rarely in a photograph and people have to go, 'oh God I didn't see you there.' That's a massive compliment in my job.
>
> *(Tony 2017)*

In contrast to the notion of invisibility that can characterise these support workers, considering aspects of their working lives that are visible offer important insights. More effective wording, however, would point to the idea of their *presence* and *absence* in particular contexts.[17] The observable activities of road crew members in the concert setting reveal details of their working lives and indicate their meaning to live music. While instrument technicians' designated place during a concert is at the side of the stage, the need for them to leave these spaces and appear onstage – which they often do hunched over to reduce visibility – means some type of technical issue has occurred that requires attention. The labours of front-of-house engineers are on display in an exclusive space in the centre of a venue's floor, surrounded by yet separated from an audience, throughout an entire show and are symbolic of their importance. Though TMs spend hours backstage sending emails to advance a show, in which their contributions to realising a concert are unseen and unknown, when they appear outside of a venue, they become the target of fans' attention (see Chapter 5). Expectations around the presence and absence of road crew members also play a role in maintaining the conventions and rituals of a concert. The "roadie cabaret," or the intermission during which crew members make final adjustments and preparations on the stage and to the equipment prior to the start of the headliner's set, serves as a marker of anticipation for audiences (Witts 2005: 147). Similarly, road crew members signify the end of a concert when the house lights are turned on and they emerge from the wings to "tear down" the stage and equipment, and during which time fans view them as opportunities to acquire a set list or drumsticks.

The significance of road crew members is represented symbolically by the system of passes (or "credentials") that define and control insider-outsider boundaries of the touring party. Passes identify and grant access to members of the touring party and label and limit the roles of those peripheral to it. The artist and road crew receive laminated passes ("laminates") that grant them "Access All Areas" or AAA for short (Reynolds 2008: 37).[18] The AAA pass allows full access to the venue, including backstage, at all times. Passes are a marker of status and prestige for outsiders, in which proximity to the artist is the allure and functions as a form of symbolic capital (see Fonarow 2006). For members of road crews, these passes are functional and allow access to the touring workplace. But in their function, they create and symbolise the road crew's access to and participation in an exclusive group.[19] Members of road crews, in particular TMs, PMs and security personnel, also reserve the power to determine who receives particular passes, including AAA laminates. They are trusted to set and control the boundaries between insiders and outsiders and shape the social composition of the backstage environment. The AAA laminate reveals that road crews simultaneously occupy positions as support personnel and have the highest, most exclusive clearance and proximity to artists. This highlights a tension in the live music hierarchy while also positioning road crews in its upper ranks.[20] In this way, AAA laminates further substantiate the notion that road crews are primary workers in supporting roles.

Understanding road crews also means grasping how they see themselves as workers in the realisation of live music events. Research respondents perceive their roles in distinction to musicians. They frame the difference in relation to industry hierarchy and musicians' centrality to live music events and their positions as artists.

> The artist is the boss, you are primarily supporting, it's all about them. Their name's on the ticket, that's an old cliché. If your name's not on the ticket then you mean nothing, you're replaceable.
>
> *(Joe 2017)*

> They're the talent, you know. It doesn't say [my name] on the ticket, you know. They're what it's all about and they're unique individuals, they're special people … And you need to treat them as special people. It doesn't mean you need to kiss their arse, you know … it's good to maintain their respect for you, but you need to understand that they are the talent, they are, I can't write a song, I couldn't stand up there and do what they do. And you need to understand that.
>
> *(Craig 2017)*

Road crew members outside of interviews and participant observation for this study further supported this view. In his guidebook on tour management, Mark Workman (2012: 315) observed that "artists of every type are unusual people and quite unlike the average person you deal with in normal everyday life." In these ways, research respondents confirm that being on a road crew involves being in a supporting role. They also recognise a tension in their roles. Both Joe (2018) and Amy (2018) see themselves as support personnel, but not exclusively as such. Joe stated that the nature of his role "depends who you're talking about. It's a support role for the artist. It's a primary role for the business, I feel like." Their roles and responsibilities place them in a position of importance to the live music industry. In other words, the live music industry could be said, in part, to depend on the services of members of road crews in realising live music, which would make them primary workers in relation to the industry. Amy described her role in terms of a professional skillset and pursuits that render the distinction between support and primary blurry.

> Kind of both. That can be a little frustrating like the further that you go up the ladder, like tour managing or whatever, cause you're doing something that you're using your organizational skills, you're using your business skills so it's a big job, but at the end of the day you're still getting paid to put your life and your pursuits of things on hold to support an artist ... I think as you go up, like production manager, tour manager, it's a little bit of both because you're kind of in that position to be primary in certain roles, but you're supporting in other parts, so it's a weird one ... you do have pride in the job that you do but sometimes it feels like your work is being, you have to put your work on hold to take care of the artist, but taking care of the artist is your job, too, does that make sense? So yeah, it's a little bit of both I think.

The fact that both of these respondents are TMs factors importantly in the perspectives and self-perception provided here. TMs have close working relationships with musicians (see Chapter 5), and the role grants them positions of relative authority at the same time that it is based on the notion of support. The last section of the chapter will discuss the topic of gender and diversity in relation to road crews.

Gender and Diversity

Road crews are marked by minimal diversity and a clear gender imbalance. The majority of workers on road crews are white men. A 2022 study by the Canadian Live Music Association titled "Closing the Gap" gave some insights into the demographics of the industry. The study covered a range of roles, but relevant here are the "workers" category which includes, but is not limited to, members of road crews. White respondents represented 35% while only 17% identified as black. The study also found that while Indigenous people and people of colour (inclusive of black people) make up 27% of the total Canadian population, they represent only 16% of live music industry workers (Canadian Live Music Association 2022: ii). During participant observation for this study, overall, representation of diverse groups of people was minimal. Of 30 artists, 6 had road crews with some visible diversity, and these crews usually worked for non-white or openly gay musicians.

The lack of diversity on road crews specifically has been covered in *Rolling Stone* magazine (Browne 2020)[21] and is evident by the presence of organisations such as Roadies of Color United[22] and Diversify the Stage[23] that work to encourage collaboration and promote a more inclusive industry. Roadies of Color United began in 2009 as a social network to "represent, introduce and unite those of us in the concert touring and entertainment services that were not being represented on other industry related social network platforms." The organisation has gained substantial following and has expanded to include a professional association that road crew members can join. The website actively promotes its members by featuring a directory of "vetted industry professionals" (Roadies of Color United 2023). Diversify the Stage (2023) is focused on ensuring that "disproportionately impacted and historically excluded communities" are included in the live music industry through "collaboration with industry leaders to actively create impactive and inclusive culture within live events, music, and touring industries." The organisation developed an apprenticeship programme to train and prepare entrants to the field and also works to establish relationships with donors and community engagement for underserved groups (ibid.). These efforts indicate the problematic underrepresentation that marks the composition of the live music industry generally, and road crews more specifically, and suggest the challenges that confront diverse groups in gaining access to the field. At the same time, they also represent and demonstrate active movement towards change.

Gender disparity is evident in the composition of the sample of research respondents and based on observations made at Venue A and Venue B. It is further addressed by organisations such as Women in Live Music[24] and SoundGirls[25] that provide resources to women interested in such a career path and raise awareness of the issues and challenges they confront. The former also includes a list of women crew members who are available to be hired for concert tours. The minority status of women on tour is further visible through various forms of media coverage that feature interviews with and stories about them.[26]

It is additionally evident by the manner in which women road crew members represent themselves when publishing their stories and offering guidance to others interested in their line of work. Kim Hawes's (2019) autobiography is called *Confessions of a Female Tour Manager*,[27] Claire Murphy's (2019) book is titled *Girl on the Road: How to Break into Touring from a Female Perspective* and Tana Douglas (2021) highlights her groundbreaking role with *Loud: A Life in Rock 'n' Roll by the World's First Female Roadie*. The specification of the authors' gender in the book titles signifies novelty, an uncommon experience and a unique perspective, and the recurrent reference to it implies a collective sense of having been unheard and unrecognised. The publishing dates of all three books – compared to men on road crews who published books a decade earlier – and the close succession of their releases reflect a longer path to being taken seriously in their roles. Women certainly do work as members of road crews, but their presence is noticeably minimal.

The self-identification of, and relative considerable attention given to, Tana Douglas as the "first female roadie" is emblematic of the gender imbalance and reinforces the exceptional and seemingly out of place nature of women on the road. It is difficult to confirm whether Douglas was actually the very first woman road crew member, and such a claim works to reproduce a mythology consistent with the conventions of wider rock culture and associated lore. However, if we are to accept this claim, being the "first" also represents a moment in which the gender composition of road crews began to change. The perception of her place on road crews as unusual and exceptional was apparent throughout her career. While working for AC/DC in the 1970s, the band's manager, Michael Browning, saw the combination of Douglas' occupation and gender as a publicity opportunity. He suggested having a newspaper article published about her that highlighted her role as a "female roadie working for AC/DC" (Douglas 2021: 78). Douglas was uncomfortable with the idea and tried to get out of doing the article. Browning argued it would be her "doing your bit for the band" – as if her job in and of itself was not sufficient – and the band members also countered her objections. Douglas was embarrassed by having her photo taken for the article and "mortified" by its headline "She Does It for the Band" (79). The band, however, found it to be "hilarious." Such divergence represents a lack of respect and support for Douglas from her colleagues that stems from her gender and related perceived expectations.

The gender imbalance is further expressed in the roles that women tend to occupy when working on tours. Gender roles are often reproduced through associations with particular positions that are considered the norm along gendered lines (see also Chapter 2). Stereotypically, women tend to occupy roles such as caterer, wardrobe assistant, merchandise vendor (see Webster 2015: 105, Vilanova and Cassidy 2019) or production coordinator more than they are found in technical or management positions. It is important to note that not all of these roles are featured on all tours and were therefore not consistently observable. Findings from participant observation were not only consistent with and substantiated the presence of the gendered nature of particular roles but also revealed some nuance. Those reported here are the exceptions. At Venue A, there was one male wardrobe assistant. At Venue B, the artist's merchandise seller was a man, and two women engineers worked on the local sound crew, one of whom was also a freelancer at Venue A and had previously worked for the local PM's company. On the local crew at Venue A, the hospitality staff and caterers were women, with the exception of two men. In management roles, there were seven women TMs and three assistant TMs present during participant observation. Two women were road managers, in the sense of the role that used to be interchangeable with the TM but now tends to focus on aspects of hospitality, and can be seen through a gendered lens. In reflecting on their years of experience, respondents noticed that the TM role is fulfilled more consistently by women than in the past and report seeing more women in technical roles than when they started working on tours. There are also several instances of artists hiring all-female road crews. However, during participant observation, women accounted for less than 10% of workers in management and technical roles.

Mary Celeste Kearney's (2017: 165) research, based on interviews in the press, indicated that ten women worked as TMs in the UK at the time.[28] Kim Hawes (2019: 184) estimated that in the first half of the 1990s, there were approximately six women TMs whom she was aware of. She further highlighted the rarity of this role in her qualification of why these women occupied these positions. She noted that four of them were "tour managers in virtue of their husbands being so" and were therefore granted access by association (ibid.). The other two had working relationships with specific bands.

Women also face challenges in management or technical positions due to their ability and qualifications to do these roles being explained and evaluated in terms of gender rather than skillset. The relative authority granted to TMs, combined with the caring features of the role (see Chapter 5), create a set of tensions that operate along gendered lines.

> I also think that women are more suited to certain roles, possibly because of a more traditional, empathic nature of women, I think they can relate to the artists a lot more … I certainly know quite a few female tour managers, I don't know if that's a mothering thing that artists require, but I certainly know a couple of female monitor engineers who, possibly as I say it's that empathic, that can relate to what people want a lot more, they're a lot more sympathetic, they're a lot more perhaps they just listen a bit more, maybe that stereotypical alpha male thing doesn't make you receptive to other people's needs as well as, perhaps that's just an idea off the top of my head.
>
> *(Adrian 2018)*

Hawes (2019: 184) was recruited as a TM by the band Concrete Blonde because of her gender. She received a phone call from a colleague who stated specifically that the band was looking for a female tour manager and that two people had put her name forward. When asked why they were looking on the basis of gender, her colleague explained that it was because the band had recently worked with, and let go, four different TMs, all of whom were male. The band's lead singer, Johnette Napolitano, asked about finding a woman for the position.

Regardless of the supposed suitability of the role, women contend with being a minority and the effects of a male-dominated workplace on their everyday working lives and experiences. Women crew members indicated a consistency in terms of their awareness of their minority status and its consideration in relation to how they navigate their careers. Due to the very limited number of women working as TMs at the time, Hawes (2019: 9) recalled that people who did not know her assumed that she would be a man. She understood herself as a "woman in a man's world" but perceived her success as related, in part, to "grit and graft" and her "refusal to conform to expectations" (2023: 4) – particularly in her rejection of becoming "one of the boys." Hawes argues that, despite some improvements across her career, it is more difficult for women to become TMs (2019: 351). Part of this is due to the nature of a managerial role and its associated responsibilities. Hawes observed that some men on road crews "don't like a woman telling them what to do" (ibid.). In turn, such behaviour creates "extra pressure to perform, to make sure that you don't slip up" (ibid.). At the same time, she also cited situations in which she felt she could use her femininity to her advantage, specifically in negotiations with male promoters.

Douglas (2021: 45) recalled that, in the process of adjusting to her new role as a crew member, the fact that she was the only woman was not immediately an obvious issue. The realisation created a sense of fear about the potential vulnerability of her position. She was concerned that any attention drawn to her minority status might result in "being thrown out of my new club." Due to this, she never asked anyone about why there were no other women working on tour. Fear of drawing attention to being a minority generated a sense of alienation and uncertainty. In contrast to Hawes' (2019) approach, Douglas described how she dressed, drank, swore and worked hard "like the boys" (45). Douglas articulated a lack of awareness of the feminist movement taking place at the time but stated that she was "striving for equality on a daily basis, without politics on my radar" (46).

As already indicated, women working in male-dominated fields and within the music industries more specifically have described experiencing the sense that they feel the need to work harder or are expected to more fully prove themselves. This can be expressed in issues related to confidence. Claire

Murphy (2019: 18) questioned her abilities and recognised her struggles with confidence on the basis of gender. She wrote about wondering whether male road crew members experience the "same type of self-doubt" and whether she had missed opportunities due to confidence issues. She also recalled the experience of working as both a tour manager and a guitar technician and, while working as the former, knew she really wanted to be the latter but "just didn't think I was good enough to be one at the highest level. Again, lacking self-confidence and doubting my abilities" (2019: 21). Issues related to women and technical roles will be further addressed in Chapter 2. The aforementioned study by the Canadian Live Music Association (2022: 50–51) also found evidence of imposter syndrome, specifically amongst women of colour, who spoke to experiencing issues related to self-confidence and feelings of unbelonging. In this way, minority identities encounter real challenges on the basis of being different from the dominant composition of the live music workplace.

Evidence of the effect of these issues can further be found in testimonies related to women's and minorities' experiences in the live music industry, which have been made visible by the UK Parliament Committee Misogyny in Music.[29] The composition of road crews is one example of a wider phenomenon. The underrepresentation of women here is consistent with a broader pattern of uneven gender representation in various workplaces in the music industries (Whiteley 1997, Clawson 1999, Leonard 2007, Hesmondhalgh 2013).[30] Relevant statistics have been reported by the UK Music 2022 Survey to Boost Diversity and Inclusion in UK Music Industry.[31] In these ways, road crews and the live music industry are not only characterised by minimal diversity and a gender imbalance – these factors have real impacts on workers. The following is a summary and conclusion of this chapter.

Conclusion

This chapter has shown that road crews are groups of specialised workers who have important individual roles and collectively participate in a shared purpose and experience. It has demonstrated that such specialisation renders the term "roadie" a problematic descriptor and inaccurate reflection of what road crew members do. With advancements in concert technology and as musicians require a diverse group to support them, crew members prefer terms that recognise their roles and contributions.

Findings presented here have offered insights into what motivates road crew members and which aspects of their work they find gratifying and have shown the support they provide to artists as among them. As behind-the-scenes workers, road crews are similar to other "invisible" workers in that their efforts are rarely acknowledged or recognised. This chapter has also suggested that the visibility of road crew members, along with their attempts to be less visible, provides important insights about their roles in live music. Such findings substantiate the significance of the activities of support personnel in their occupational identities as well as in the realisation of live music.

Road crew members destabilise notions of hierarchy in the music industries by simultaneously being support personnel and having the highest level of access in the backstage environment. They also complicate clear-cut understandings of the secondary status of support personnel by self-identifying in distinction to artists and recognising their central importance to the live music industry.

This chapter has also shown that road crews are marked by an underrepresentation of women and minimal diversity, which means they reproduce exclusionary patterns that exist in the broader music industries. Women's positions on road crews can be and are influenced by stereotypical expectations of gender norms and roles, and their experiences are often shaped and permeated by their gender and its minority status in touring work. The next chapter will explore and examine the features of gaining and maintaining work as a member of a road crew.

Notes

1 Available at these links: https://www.rollingstone.com/music/music-features/music-in-crisis-marty-horn-tour-manager-rolling-stones-stevie-nicks-995835; https://www.rollingstone.com/pro/features/roadie-clinic-live-music-touring-advocacy-1035731; https://www.rollingstone.com/music/music-news/covid-19-music-industry-charities-help-1122999/
2 Available at these links: https://www.billboard.com/articles/business/touring/9340968/coronavirus-concert-cancellations-touring-gig-economy; https://www.billboard.com/music/music-news/live-nation-crew-nation-coronavirus-fund-9348011/; https://www.billboard.com/articles/business/touring/9357092/protect-touring-crews-coronavirus-over-tom-ross
3 Available at these links: https://www.nytimes.com/2020/03/17/arts/music/concerts-cancelled-coronavirus.html
4 Available at these links: https://www.theguardian.com/music/2020/mar/21/coronavirus-leaves-roadies-and-events-crew-devastated-its-the-first-industry-to-stop-dead; https://www.theguardian.com/music/2020/aug/15/the-terrible-plight-of-music-event-staff-coronavirus-pandemic?mc_cid=21913aa5d5&mc_eid=305aa65491; https://www.theguardian.com/music/2020/dec/04/one-two-one-two-the-scottish-bands-teaming-up-to-support-roadies
5 *Roadie*, 2011.
6 *Roadies*, Showtime, 2016.
7 "Roadie," Tenacious D, 2012, "Roadie Man," Pretenders, 2016.
8 Chapter 2 will show how working for friends is an important part of learning and training for members of road crews.
9 Gorman (1978: 25–28) also countered the perspective held by "casual observers of the rock scene" who "seem to believe that roadies are basically unskilled people…"
10 https://www.backstageculture.com/roadie-dictionary-a-list-of-touring-terms/

11 See, for example, the book *Hammer of the Gods* by Stephen Davis ([1985] 2008) or the Rolling Stones documentary *Cocksucker Blues* (dir. Frank 1979).
12 Podcasts and merchandise available at roadiefreeradio.com
13 See https://wholelottaroadies.bandcamp.com/album/whole-lotta-roadies
14 Website is roady4roadies.com
15 See the roadieclinic.com
16 https://www.instagram.com/theroadiecookbook. Also available at theroadiecookbook.com
17 See Webster (2011: 91–92) for a discussion on covertness and overtness in relation to the role of concert promoters and their presence at live music events.
18 People affiliated with the touring party may also receive AAA laminates, such as record label executives, family and booking agents. They may have full access or it may be limited by restricting escort privileges; in the latter case, a small sticker would be added to the laminate (Reynolds 2008: 39). Other passes that indicate specific levels of access include Guest, Aftershow and Working. The Guest pass allows an individual to go backstage after the show and to remain within the venue after the show is over. The Aftershow pass allows a person to remain in the venue after the audience leaves and to participate in aftershow activities, usually a party. The Working pass is given to members of local crews (see Chapter 3) or other people working on a show day. These passes are usually stick-on and are differentiated by colour. Venue security personnel receive a document that features copies of and details about the passes on a given show day, which is also posted at access points (ibid.).
19 In her study of British indie musicians, Wendy Fonarow (2006: 147) observed that road crew members formerly used laminates as representation of their career histories. They would wear all of their previous laminates as a "portable resume" which other crew members could view to "determine what colleagues they shared in common." Fonarow noted that though this practice was common in the early 1990s, it was "uncool" by the 2000s and is now seen as "lame." Furthermore, she described how a crew member's current band will make fun of the other bands that the crew member has worked with. None of these practices were observed during participant observation or reported by research respondents during research for this book.
20 Webster (2011: 131) further substantiated the place of road crew members in the live music hierarchy when observing that the TM in particular is "regarded as the 'top of the tree' and is the person ultimately responsible for the entire tour once it is on the road, in terms of looking after and managing the crew."
21 https://www.rollingstone.com/music/music-features/music-road-crews-diversify-the-stage-1097096/
22 http://www.rocu.tours/
23 https://www.diversifythestage.org/
24 https://womeninlivemusic.eu/
25 https://soundgirls.org/
26 See *The Guardian*: https://www.theguardian.com/music/2007/apr/13/popandrock.gender; NPR: https://www.npr.org/2016/09/04/492433224/meet-the-woman-whos-been-pearl-jams-sound-engineer-for-24-years
27 Hawes released another book in 2023 titled *Lipstick and Leather: On the Road with the World's Most Notorious Rock Stars*.
28 See https://www.independent.co.uk/arts-entertainment/music/features/where-are-women-rock-music-industry-2009515.html
29 https://committees.parliament.uk/event/14556/formal-meeting-oral-evidencesession/
30 Efforts have been and continue to be made to call attention to and change the gender composition of various sectors of the music industries. See UK Music's

(2020) diversity report: https://www.ukmusic.org/equality-diversity/uk-music-diversity-report-2020
31 https://www.ukmusic.org/equality-diversity/uk-music-diversity-report-2022/

References

Alacovska, Ana and Joëlle Bissonnette. 2019. "Care-ful Work: An Ethics of Care Approach to Contingent Labour in the Creative Industries." *Journal of Business Ethics* 169: 135–151.
Ames, Richard. 2019. *Live Music Production: Interviews with UK Pioneers*. New York: Routledge.
Arrow, Holly and Joseph E. McGrath. 1995. "Membership Dynamics in Groups at Work: A Theoretical Framework." *Research in Organizational Behavior* 17(1): 373–411.
Banks, Mark. 2007. *The Politics of Cultural Work*. Basingstoke: Palgrave.
Battentier, Andy. 2021. *A Sociology of Sound Technicians: Making the Show Go On*. Wiesbaden: Springer VS.
Becker, Howard S. 1982. *Art Worlds*. Berkeley: University of California Press.
Bennett, H. Stith. (1980) 2017. *On Becoming a Rock Musician*. New York: Columbia University Press.
Browne, David. 2020. "Music's Road Crews Are Overwhelmingly White and Male. Meet the People Trying to Change That." *Rolling Stone* (17 December): https://www.rollingstone.com/music/music-features/music-road-crews-diversify-the-stage-1097096.
Canadian Live Music Association. 2022. "Closing the Gap: Impact & Representation of Indigenous, Black, and People of Colour Live Music Workers in Canada." https://www.canadianlivemusic.ca/closing-the-gap#:~:text=Our%20Purpose,its%20project%20partners%20and%20funders.
Clawson, Mary Ann. 1999. "Masculinity and Skill Acquisition in the Adolescent Rock Band." *Popular Music* 18(1): 99–114.
Cunningham, Mark. 1999. *Live and Kicking: The Rock Concert Industry in the Nineties*. London: Sanctuary Publishing.
Curtin, Michael and Kevin Sanson, eds. 2017. *Voices of Labor: Creativity, Craft, and Conflict in Global Hollywood*. Oakland: University of California Press.
Davis, Stephen. (1985) 2008. *Hammer of the Gods: Led Zeppelin Unauthorized*. London: Pan Books.
Diversify the Stage. 2023. Official Website. https://www.diversifythestage.org.
Douglas, Tana. 2021. *Loud: A Life in Rock 'N' Roll by the World's First Female Roadie*. Sydney: ABC Books, HarperCollins Publishers.
Eliot, Marc. 1989. *Rockonomics: The Money Behind the Music*. London: Watts.
Fonarow, Wendy. 2006. *Empire of Dirt: The Aesthetics and Rituals of British Indie Music*. Middletown, CT: Wesleyan University Press.
Frank, Robert, dir. 1979. *Cocksucker Blues*. Video Beat.
Frith, Simon, Matt Brennan, Martin Cloonan and Emma Webster. 2021. *The History of Live Music in Britain, Volume III, 1985–2015: From Live Aid to Live Nation*. London: Routledge.
Ginnett, Robert C. 2019. "Crews as Groups: Their Formation and their Leadership." In *Crew Resource Management*, eds. Barbara Kanki, Robert Helmreich and José Anca, 73–102. San Diego, CA: Academic Press.

Gorman, Clem. 1978. *Backstage Rock: Behind the Scene with the Bands*. London: Pan Books.
Hart, Colin. 2011. *A Hart Life*. Bedford: Wymer Publishing.
Hawes, Kim. 2019. *Confessions of a Female Tour Manager*. Independently Published.
Hawes, Kim. 2023. *Lipstick and Leather: On the Road with the World's Most Notorious Rock Stars*. Muir of Ord, Scotland: Sandstone Press, Ltd.
Hesmondhalgh, David. 2013. *The Cultural Industries*. 3rd ed. London: Sage.
Hesmondhalgh, David and Sarah Baker. 2011. *Creative Labour: Media Work in Three Cultural Industries*. New York: Routledge.
Hince, Peter. 2011. *Queen Unseen: My Life with the Greatest Rock Band of the 20th Century*. London: John Blake Publishing.
Kearney, Mary Celeste. 2017. *Gender and Rock*. Oxford: Oxford University Press.
Klimoski, Richard and Robert G. Jones. 1995. "Staffing for Effective Group Decision Making." In *Team Effectiveness and Decision Making*, eds. Richard A. Guzzo and Eduardo Salas, 9–45. San Francisco, CA: Jossey-Bass Publishers.
Leonard, Marion. 2007. *Gender in the Music Industry: Rock, Discourse and Girl Power*. Aldershot: Ashgate Publishing Limited.
Leyshon, Andrew. 2009. "The Software Slump?: Digital Music, the Democratisation of Technology, and the Decline of the Recording Studio Sector within the Musical Economy." *Environment and Planning A* 41(6): 1309–1331.
Mayer, Vicki, Miranda J. Banks and John T. Caldwell, eds. 2009. *Production Studies: Cultural Studies of Media Industries*. New York: Routledge.
McGinn, Matt. 2010. *Roadie: My Life on the Road with Coldplay*. London: Portico Books.
Murphy, Claire. 2019. *Girl on the Road: How to Break Into Touring from a Female Perspective*. Independently Published.
Oxford English Dictionary (OED). 2020. s.v. "Support," accessed September 18, 2020, https://www.lexico.com/en/definition/support.
Reynolds, Andy. 2008. *The Tour Book: How to Get Your Music on the Road*. Boston, MA: Cengage Learning.
Reynolds, Andy. 2012. *Roadie, Inc.: How to Gain and Keep a Career in the Live Music Business*. 2nd ed. CreateSpace Independent Publishing Platform.
Reynolds, Andy. 2020. "What Does a Tour Manager Do?" *Live Music Business*. https://livemusicbusiness.com/artist-resources/what-does-a-tour-manager-do.
Roadies of Color United. 2023. Official Website. https://www.rocu.tours.
Shuker, Roy. 2008. *Understanding Popular Music Culture*. 3rd ed. London: Routledge.
Shuker, Roy. 2013. *Understanding Popular Music Culture*. 4th ed. London: Routledge.
Simpson, Dave. 2009. "Roadie to Nowhere: Why Time Can't Change the Rock Road Crew." *The Guardian* (13 April): https://www.theguardian.com/music/musicblog/2009/apr/13/roadie-rock-road-crew.
Sullivan, Caroline. 2014. "The End of the Roadie: How the Backstage Boys Grew Up." *The Guardian* (12 June): https://www.theguardian.com/music/2014/jun/12/end-of-the-roadie-techs-live-music.
Sundstrom, Eric, Kenneth P. DeMeuse and David Futrell. 1990. "Work Teams: Applications and Effectiveness." *American Psychologist* 45(2): 120–133.
Thomas, Jawsh. 2020. "Roadie Dictionary: A List of Touring Terms." *Backstage Culture*. https://www.backstageculture.com/roadie-dictionary-a-list-of-touring-terms.
Tronto, Joan C. 1993. *Moral Boundaries: A Political Argument for an Ethic of Care*. New York: Routledge.

Vilanova, John and Kyle Cassidy. 2019. "'I'm Not the Drummer's Girlfriend': Merch Girls, Tour's Misogynist Mythos, and the Gendered Dynamics of Live Music's Backline Labor." *Journal of Popular Music Studies* 31(2): 85–106.

Virtue, Graeme. 2012. "One for the Roadies: In Praise of Rock 'n' Roll's Unsung Heroes." *The Guardian* (12 December): https://www.theguardian.com/music/2012/dec/12/roadies-rock-roll-unsung-heroes.

Webber, Sheila Simsarian and Richard J. Klimoski. 2004. "Crews: A Distinct Type of Work Team." *Journal of Business and Psychology* 1(3): 261–279.

Webster, Emma. 2011. "Promoting Live Music in the UK: A Behind-the-Scenes Ethnography." PhD diss., University of Glasgow.

Webster, Emma. 2015. "'Roll Up and Shine': A Case Study of Stereophonics at Glasgow's SECC Arena." In *The Arena Concert: Music, Media and Mass Entertainment*, eds. Benjamin Halligan, Kristy Fairclough, Robert Edgar and Nicola Spelman, 99–109. New York: Bloomsbury Academic.

Weinstein, Deena. 2003. "Roadies." In *Continuum Encyclopedia of Popular Music of the World, Part 1: Media, Industry, Society*, eds. John Shepherd, David Horn, Dave Laing, Paul Oliver and Peter Wicke, 565. London: Bloomsbury Publishing.

Whiteley, Sheila, ed. 1997. *Sexing the Groove: Popular Music and Gender*. London: Routledge.

Witts, Richard. 2005. "I'm Waiting for the Band: Protraction and Provocation at Rock Concerts." *Popular Music* 24(1): 147–152.

Workman, Mark. 2012. *One for the Road: How to Be a Tour Manager*. Road Crew Books.

Wright, James "Tappy" and Rod Weinberg. 2009. *Rock Roadie: Backstage and Confidential with Hendrix, Elvis, The Animals, Tina Turner and an All-Star Cast*. London: JR Books.

Umney, Charles and Lefteris Krestos. 2015. "'That's the Experience': Passion, Work Precarity, and Life Transitions among London Jazz Musicians." *Work and Occupations* 42(3): 313–334.

Interviews

Adrian. Guitar Technician. In-Person Interview, 24 May 2018.

Amy. Tour Manager. Skype Interview, 9 July 2018.

Craig. Monitor Engineer. In-Person Interview, 4 May 2017.

Duncan. Guitar Technician. Phone Interview, 10 May 2018.

Joe. Tour and Production Manager. In-Person Interview, 2 May 2017, 14 May 2018.

Tony. Close-Protection Security, Head of Security. Skype Interviews, 4 August 2017, 16 March 2018.

2

GETTING IN, GETTING HIRED, WORKING, LEAVING

The path to becoming a member of a road crew is marked by variation and informality and formed by the spaces and people crew members work in and amongst. It is not standardised or marked by formal stages of progression or professionalisation. This chapter is about the norms and practices that shape the career paths of road crew members and shows how they become part of the live music industry. It discusses the ways in which road crew members acquire their skillsets, learn the responsibilities and expectations associated with their roles and how to fulfil them and the types of training that are available. The chapter explores the manner in which road crew members gain access to the live music industry and provides an analysis of the ways in which they maintain work. It also considers issues related to getting fired and reasons crew members may choose to leave the road. The chapter demonstrates that the most important factor in gaining and maintaining work is a strong reliance on a network of contacts and highlights that road crew members navigate their working lives through a set of informal practices.

Learning and Training

For future members of road crews, music generally plays an important formative role but is not characteristic of all of them. Members of road crews often have shown a strong interest in music from early in their lives. With the exception of Tony (2017, 2018) whose motivation to work in the live music industry was part of a broader career trajectory in close protection security, all respondents spoke of an interest in and enthusiasm for music that began at a young age. The motivating factor of music is further evident in the books and autobiographies that members of road crews have published (e.g., Wright and

DOI: 10.4324/9781003303046-3

Weinberg 2009, McGinn 2010, Hawes 2019, Murphy 2019, Douglas 2021) and in other interviews with crew members.[1]

Scholars have found that a successful career in live music develops despite an absence of "strict rules, formulas or entry requirements" (Frith et al. 2013: 18). Consistent with this observation, there is no formal qualification to become a member of a road crew. As Joe (2017) summarised, "most people I know in the same circumstances have come into it in the same way and I think fundamentally that's without any official training." The typical site and method for learning is on the job and through practical experience. Crew members initially become involved in working in live music in their local area by working for promoters and at venues or university student unions. Amy (2018) described how she "grew up as a production runner and worked for production companies" that gave her experience with and access to tours coming through venues in the local area. Joe (2017) spent "a lot of time in the venues" working behind the bar, helping his friends' bands, assisting with organising local nights and setting up the PA. The student union at Craig's (2017) UK university featured prominently on national touring circuits at the time. Initially motivated by the opportunity to gain free admission, he worked for the union by hanging posters up around the city to advertise for concerts. Doing so turned into a paying job on the stage crew, which he described as a local "training ground for professionals" as it comprised very few students.

UK student unions have been significant to live music and its workers since the mid-1950s and started to become important venues on the touring circuit by the mid-1960s, particularly for the jazz and folk genres (Frith et al. 2013: 179–180, 2019: 63–64). During this time, the union role of "social secretary" became an established position. It involved booking music and other forms of entertainment and provided an opportunity to work with agents, negotiate fees and produce live events for large student audiences. It was likewise identified as an important starting point for commercial promoters in particular and was a "new path into the live music business" by the mid-1960s (ibid.). Craig noted that the majority of workers from his student union's crew are still in the live music industry in some capacity. In this way, research findings in this study indicate that student unions have functioned as important training sites and a "path into the live music business" for a *range* of workers in live music, including future members of road crews.

As mentioned in Chapter 1, an early job for crew members is helping friends who are musicians. This usually involves driving friends to and from gigs and assisting with moving equipment. Adrian (2018) summarised the experience of working for his friends' bands.

> And that took kind of like a four-year period that was probably like, in retrospect, a training ground for some of the basic skills, although it didn't feel like that at the time, it was just an excuse for traveling around with my

> mates having a nice time. But yeah, in hindsight, you get a lot of the basic skills there, the basic skills really being picking stuff up and carrying them and you know learning how to keep to schedule and having a little bit of discipline around schedule and tasks and things. But it was all very very informal.

As Adrian's account suggests, the act of helping friends is not necessarily taken, at the time, as a serious step towards a career path, and the value of early and informal experiences does not become apparent to crew members until later. Joe (2017) similarly reflected on his early experiences on tour as having been characterised by driving a van and selling t-shirts for friends' bands "just helping out ... for fun a couple times." Duncan (2018) echoed this when explaining that he lived in a shared house with a musician who played "some odd gigs around the clubs of the UK and stuff ... he said, 'do you want to come along and help out for some beers and a laugh?' and I said, 'oh yes please'."[2] Claire Murphy (2019: 40) also noted that bands will have their friends drive the van to "help them out for 'beer money'." The informal manner in which they take on these roles, and the pleasure they assign to them, is foregrounded over any association with work. However, these contacts and experiences create opportunities to develop skills and gain exposure to the field.

Skills are further developed by observing, working with and learning from more experienced co-workers through instruction and mentorship. Venues facilitate regular access to incoming touring parties, which creates contact with a wide range of skilled workers from whom future road crew members can learn. Research respondents described talking with visiting crew members about their jobs in order to learn as much as possible about their roles. While venues are resources for on-the-job training, crew members in turn demonstrate high levels of initiative and commitment by taking advantage of and seeking out learning opportunities. The importance of venues in this context reinforces their significance and highlights further problems with the threat to the existence of smaller ones. The numerous closures of small, grassroots music venues, and uncertainty over their survival, are a major focus of concern, given their role in facilitating development and exposure for upcoming musicians.[3] However, these spaces also serve as sites of learning, development and networking for support personnel. As such, small venues are equally important to many types of workers, and issues pertaining to them have an impact on a range of careers in the live music industry.

Some members of road crews were musicians themselves and moved from intending or attempting to make a living as one to working on a crew. Being a musician facilitated direct participation with the live music industry and offered important transferrable skills. For those crew members, their interest in music and the realities of being a musician gave way to needing to decide, or realising, what to do with their career. Adrian (2018) described this process.

> There comes a time when you have a realisation that you might not have either the talent or the opportunities or just the plain look to make it in the music business as a musician yourself. Certainly in my case after five years of trying I came to that conclusion. I was getting older as well and the windows of opportunity as a musician were all closing very very rapidly. And it's kind of like when you take your stock of yourself as a person and what skills you have, particularly skills [that] might be transferrable to the general job market and I didn't really have very many. The only thing that I knew about in any depth was messing around with guitars and amplifiers and you know various bits of audio gear. And it became, 'well, how can I turn what I know and what I'm barely good at into an on-going career?' So becoming a technician in the concert industry was possibly the best opportunity to stay within the music business itself but within a different role in a different area in it.

Crew members who were musicians also come to realise that they are better suited and more interested in working in different roles. Being a musician leads them to discover this line of work.

> I was in a band … I wasn't in a particularly good band, and I wasn't a particularly good bass player. But it led me into this and I found out I was better at this, I like this better. I don't want to be that guy. Do you know what, it's too much hanging out if you're a musician. All day for 90 minutes. I think I would develop a drinking or a drug problem fairly quickly, I'd just get bored. Whereas the tour manager or the production manager are effectively the two busiest people on the road. You get no point at which in the day when somebody might not come and ask you to solve a problem. I like that.
>
> *(Joe 2018)*

Such road crew members are not simply "failed musicians," which implies that the latter is considered the more desirable role and becoming a crew member is a secondary decision that they simply settle for. Rather, there are both push and pull factors in their shift to working on a crew, and they locate suitability and legitimacy in their positions.

Participation and enthusiasm for live music can also help crew members identify their path. In her book on touring, guitar technician Claire Murphy (2019: 9) described how her initial "passion for live music" was also connected to a desire to form a band and tour. She eventually realised that her interest had less to do with being in a band and more to do with going on tour, which prompted her to seek work as a crew member.

Research findings indicate that though road crew members eventually specialise, they tend to learn multiple roles early on, and that doing so allows them to gain experience and opportunities that help them navigate the field (see Reynolds 2008: 382–384). The ability to draw on a variety of skills is useful for job entry and can be an asset for continued employment. In this way, they are like other professionals in the live music industry in that they often wear "several hats at once, a form of multi-tasking that could be said to define this business sector" (Frith et al. 2013: 18). It is important to note that scholars have identified this tendency in terms of the same person occupying the role of "manager, agent, promoter, venue owner and, indeed, musician" (ibid.). This pattern amongst support personnel not only identifies consistency amongst workers in live music but also highlights the range of, and reinforces the distinction between, specialised roles on road crews that serve important functions in the industry. Amy (2018) strongly emphasised the importance of learning multiple roles, encouraging potential entrants to "do like do everything you can … you will learn everything." Kim Hawes (2019: 346) echoed this statement and positioned its value specifically in relation to developing the necessary skillset to be a tour manager. Having a working knowledge of all the roles on tour equips potential TMs with information that will assist them in their everyday lives so that they "can see that everything gets done that needs to get done." Doing so offers opportunities to understand the wider division of labour in live music at the same time it can help with the development of a specialism.

> At that time I was assistant carpenter. The beauty of the Roundhouse was that we were all able to do something of everything. My job was carpenter but effectively I used to do work on the doors, on security, behind the bar sometimes, in the catering area but my full-time job was carpenter for the place. That developed into eventually becoming production manager for the Roundhouse.
>
> *(Ames 2019: 36)*

Making an effort to learn a variety of roles also widens the scope of potential work opportunities. Duncan (2018), who now works exclusively as a guitar technician, started out by working multiple roles. Being a "bit of a jack of all trades" was beneficial for how it reduced the risk of being bypassed for a job. Learning multiple roles makes a crew member "indispensable" and "will get you hired on more tours," which helped Claire Murphy (2019: 20) early in her career by doubling as a tour manager and guitar technician. Road crew members also cultivate opportunities by being available, which is another key factor in both learning and acquiring work. Research respondents reported that being present and taking every job that was offered enhanced learning and access. This sometimes involved a willingness to make themselves available or rearrange their schedules even if they were already booked.[4] It can also involve taking risks to create space for opportunities. After making a connection at a live rehearsal studio that offered the possibility of a few shifts, Murphy (2019: 14–17) quit a full-time, well-paid position and took another job that allowed her the flexibility to make herself available to the studio. However, it led to a more permanent, albeit part-time role that put her in contact with the person who helped her learn and move successfully into touring. Working at venues, assisting friends, learning multiple roles and being available are the ways that crew members "pay their dues" and prove themselves in the early stages of their careers.

The ability to learn multiple roles and adapt to a variety of jobs was also a useful skill and resource for road crew members during the COVID-19 pandemic. The complete shutdown of live music events meant crew members – arguably one of the groups of workers that were most effected – needed to find alternate ways of earning a living during lockdown. As example, Bell, Bell and Taylor (2022: 1938) recounted the experience of Chad, a musician and "a jack-of-all-trades when it comes to staging a live performance."

> Like most gifted people in the industry, he is versatile and is able to nimbly adapt to changed situations. Currently, to keep body and soul together, he is working for moving van and cartage companies because he cannot load/unload instruments and props now that there are no live shows. As a technician, Chad has been able to transfer his skills to another employment sector—at least temporarily.

Crew members were able to temporarily "ply their trades outside of the venue setting" in transferrable roles and adapt to a "'new normal' in a variety of creative ways" (1928–1929). These included jobs such as delivery drivers in various contexts and related positions in other sectors of the cultural industries.

For women, gender can be a factor in the types of roles they are able to learn and pursue. As within the wider music industries, the "predominance of men working in the industry has created problems of access and opportunity for women" interested in working in the live music sector (Leonard 2017 [2007]: 24). Tasha (2018) has held multiple roles as a crew member. Gender factored significantly into the path she took based on how it inhibited access to training.

> I think that when I started I had a keen interest in sound design, but there just weren't any women doing it and I had no mentors, I didn't have anybody I could talk to about it and the idea at the time of somebody who's kind of a girly girl doing sound was just so preposterous that I would have front of house guys humour me and be like 'yeah during the show come up here and you can watch what I'm doing,' but no one would ever take me under their wing and teach me anything and if I have any major regrets,

> that's what I wanted to do but it was too hard for me to even get into it or find someone to mentor me, I just decided to stick with doing the admin jobs because it was where I could get work, and doing merch or doing wardrobe, because these were the soft-touch jobs that I could get, that would keep me on the road.

Tasha's story shows the limitations and frustrations women encounter in trying to access male-dominated roles. While trying to break into touring, Murphy (2019: 12–13) worked in a guitar shop and was told by a male colleague that "women shouldn't work in music and should not be allowed to play guitar." Years later, she was given an opportunity to shadow a guitar technician on tour, which helped her understand her role in depth. Due to the experience, Murphy stated that she believes mentorship is important and offers the chance for aspiring women to shadow her (17). When initially realising the relative precarity and uncertainty of working on tour, Tana Douglas (2021: 48) became concerned about the likelihood of being hired by the friends she had made on road crews. She questioned whether they would choose her over "their old mates" and "over another guy" (ibid.).

Women who are able to train and acquire work as sound technicians can also encounter difficulties on the basis of gender (Kearney 2017: 159), which can greatly shape and affect their everyday work experiences. The male-dominated workplace and masculine culture of live music and touring (see Chapter 4) affect women's ability to be taken seriously in their everyday working lives. Similar to experiences of women working in other sectors of the music industries, such as electric guitarists and recording studio engineers, women crew members are often misidentified on the basis of gender. In a study by Smaill (2005: 16, Leonard 2017 [2007]: 58), a live sound engineer described being denied access to a venue because the "staff did not believe that a woman could be operating the sound desk." Smaill also noted that comparable experiences had been reported by other women working in the same role. Such an example demonstrates the extent of the influence that the masculinist culture, which derives from the male-dominated composition of road crews, has in the live music setting. It is an instance "of confirmation, where the backstage area is re-established as a male space" (Leonard 2017 [2007]: 58). Douglas (2021: 65), who worked as both a backline technician and a FoH engineer during her career, recounted a similar experience. While working at a music festival with AC/DC in the 1970s, she made her way to the stage in order to begin time-sensitive preparations and was stopped by a security guard who told her "You can't come back here, little lady." When she assured him that she was with the band, he directed her to locate the members because he believed she could not be allowed on the stage. After clarifying that she was a crew member, the security guard stepped towards her and stated that he "won't tell you again!" It was only when a male crew member already on stage yelled to the security guard that the information Douglas was providing was "for real! She's their roadie and you have to let her up here" that she was allowed to pass. Douglas reflected how her "word meant nothing" and that such occurrences were "often" during which she would have to convince people of the reasons for her presence and that she was not there in the pursuit of musicians (65).

Early in her career, Douglas (2021: 97) believed that career advancement would be based on "how well I fitted in as a crew member, not how I stood out as a girl." In other words, the exceptional nature of being a woman on a crew was seen as a potential deterrent. Her perspective was based on continuing difficulties she encountered on the basis of gender and the attitudes and beliefs expressed by her male counterparts about her abilities. She recalled male crew members making jokes or pushing her "out of the way when there was a heavy lift" (ibid.). They would instruct her to move aside and let the male crew members handle such tasks as they did not want Douglas to be injured (ibid.). She concluded that this behaviour functioned as a "frequent reminder that there was a group who didn't want me there" (ibid.). As such, her gender factored into her ability to be taken seriously and had the potential to impact her future career.

While working for promoter Paul Dainty in Australia, Douglas recounted how the crew's name was "jokingly" changed from the Paul Dainty Road Crew to the Very Dainty Road Crew in "honour of having a girl on board" (105). Being seen simultaneously as a humorous and validating act, the renaming problematically reaffirmed Douglas' exceptional nature while drawing on and reproducing gender stereotypes that could potentially undermine her abilities. At the same time, Douglas also became aware that the male members on her crew held her in respect and would not "mess with her" because they were certain she "could hold up to the best of them" (125). Where Douglas did achieve some sense of normalcy, however, was when she was hired to work for Suzi Quatro. As both were "unusual young women at the time" she "fit in well with her and her crew" and Quatro "liked the fact that she has a girl on the crew" (105–106). In other words, the two women experienced a sense of solidarity and shared experience on the basis of their marginal status in the music industries, which in turn created a cohesive working environment.

These accounts are consistent with a longer history of problems that women contend with that are related to the intersection of gender, technology and learning in the music industries (e.g., Bayton 1997, Bourdage 2010). The pervasive view of technology as "masculine" fosters the exclusion of women. Informal training and mentorship mean that male territorialism around technology can act as a barrier to knowledge transmission for women.

The difficulty of getting in can also be understood in gendered terms. The unequal gender representation can be linked to an unconscious bias that can affect women's ability to gain entry to the field and acquire jobs. Joe (2017) recalled initially becoming aware of the gender imbalance when he was working for a promoter, prior to the start of his touring career. There, he described how women tended to reach a certain level but had difficulty advancing further. Women were rarely hired for any of the company's large concerts, but "boys got the big shows." During interviews, Joe substantiated that this can and does continue to occur. In his capacity as a TM and PM, he tries to hire women whenever possible and explained that "when I do come across women working in the business [I try] to not be a barrier to them, to encourage them, to put down misogyny when I hear it" (2019). This perspective and approach highlight the importance of awareness and show that hiring practices can be obstacles for women – or can be removed. Women interviewed for this study did not universally feel discriminated against or disadvantaged in their careers on the basis of gender and also cited that men played an integral role in assisting them with access to work. Findings therefore suggest that men can both inhibit and facilitate women's opportunities for learning, access to work and advancement on road crews.

It is here that it is important to remind the reader that this study is focused on a particular demographic of crew member, those that have an average of 20 or more years in the industry and began working at a particular time. While the informal practices of on-the-job-training and assisting friends are still utilised and accepted in terms of learning and acquiring work, there are now alternative means and a diverse set of possibilities in the form of college and university programmes. These include the Academy of Live Music Technology,[5] run by former TM Glen Rowe, and the Technical Services Route programme at The Academy of Contemporary Music in partnership with Middlesex University London.[6] The former offers both undergraduate and postgraduate degrees in areas including Live Events Production and Live Visual Design, while the latter is a series of modules, including Live Event Management and Live Sound, that are taken as a component of an undergraduate programme in Music Industry Practice. In addition to these programmes, the annual event Production Futures,[7] held in the UK, is a convention and networking event designed to assist job seekers. While it is unknown to what extent, or even if, these training programmes will shape the industry and its practices over the long term, it is important to highlight that these are options that were largely unavailable to the group of workers in this study. As noted in 2007, "few, if any, colleges have ever offered comprehensive educational programs to train individuals for careers in the concert industry, anyone working in this business has had to either learn their craft through on-the-job training or via fairly well-structured mentoring systems" (Waddell, Barnet and Berry 2007: 2,

Reynolds 2008: 381). There is evidence of prior interest in developing a standardised or formal system of professionalisation, especially in consideration of advancements in concert technology. In the 1960s, Claude, the "head roadie" for the band Family, observed that crew members had to attend trade fairs in order to keep up with technology. He endeavoured to start an agency to train and advise younger workers and believed that they would "do pretty well" (see Frith et al. 2019: 73). A similar observation was made nearly 40 years later in 2004 by Feargal Sharkey, formerly of The Undertones, and then-chair of the UK Government's Live Music Forum, who noted that the live music scene was in "great shape" but suffered from "a lack of trained technicians" (Frith 2007: 3).

The relative success rate of these programmes in facilitating entry into the industry, and the acceptability of these programmes in the live music industry, is as yet not fully clear or known. However, road crew members do express some hesitation and even resistance towards them. TM Kim Hawes (2019: 345) stated that a university course is not "enough" to work as a tour manager and argued that "experience is essential." Duncan (2018) found it difficult to connect classroom training to the actual experience of live music.

> ... you can learn all the theory but you've still gotta have that hands-on training to get live show experience. And the pressure is there whether you've got 50 people heckling you or 50,000 people heckling you ... so you know you can sit in the classroom and be taught how things go together or where to plug things in but I don't quite understand how it works in terms of then going off to do shows for the band.

Joe (2017, 2019), who taught related music business courses while he was recovering from an injury and could not tour, does not see such programmes as viable pathways to careers in live music. In his view, they provide useful knowledge, access to experienced people in the field and may help students recognise which particular area of the industry is of interest. However, they do not compensate for the hands-on experience and contacts made working in local venues, nor do they guarantee students jobs in live music. Such a qualification is "useless" and is not what industry personnel are looking for, though it does have the merit of demonstrating a certain level of commitment over a two-year period. The "good students" take advantage of the information available to them in the course and simultaneously work in a bar or live music venue to gain experience – and may leave the course once they have gained enough information. He criticised the courses for the ways in which he perceives them as taking advantage of "a lot of people who want to be on the road and taking their money" (2017). They are "taking people's money and giving them false hope, giving them some idea that this is gonna lead somewhere. It absolutely doesn't (2019)." Simon Frith (2007: 13) similarly questioned the continued attraction of conservatories and stage schools despite the fact that the majority of "graduates will not make a living from performing." Sally Anne Gross and George Musgrave (2020: 137) have also questioned the value and approach of such courses and the "same dream" being sold and perpetuated with them. They referred to the practice of "cramming more and more students" into these programmes without rethinking the built-in and taken-for-granted promotion of the music industries' "mythological vision" of itself as irresponsible (ibid.). For crew members who no longer want to tour as often or at all, however, the courses provide, and are regarded as offering, access to employment as teachers. Such statements from research respondents represent an ambivalence regarding the courses and crew members' own participation in them.

An off-the-record conversation with a panel participant at a Production Futures event further confirmed the predominantly negative view of university training courses. This participant understood the technical programmes as primarily being a product of how quickly concert technology changes and develops. Despite this, he does not believe that the informal, largely word-of-mouth mode of recruitment is completely changing – or that it should. He explained that if he was given 20 CVs and one of them was a referral from someone he knew, he would hire that person over someone with a degree. He added that he thinks students have no idea what working on tour is actually like; they see the glamour, but have no idea how hard it is, and highlighted that crew members are "lucky" if they have an 18-hour day with little sleep and are crammed into a bunk or the back of a van. Observations of attendees at the Production Futures event validated some of these statements. Potential entrants to the field were overheard commenting on the attractive features of the interior of a tour bus, which was onsite for attendees to explore, and expressing excitement over its array of amenities, such as a television and PlayStation.

The previous comments by road crew members reflect an adherence to accepted practices and some mistrust towards these university courses, which is based primarily on the disconnect between the limits and possibilities of classroom learning with real-world experience. They also highlight the importance of the culture and experience of touring in being a suitable worker and in understanding what that entails. As this book will show in more detail, a key component in working on tour involves participating in its social norms and culture. In their resistance to such types of tertiary education, their comments reinforce that their informal practices have come to be taken as the norm or standard. This is similar to reactions and perspectives from workers in recording studios following the development of similar training programmes for their field. Susan Schmidt Horning (2004: 719) observed that studio engineering continued to "value on-the-job training over formal education, even after such education had become available." This was based on a "general attitude within industry that schooling can take one only so far, and that the real training is up to the employer and occurs in the workplace" (ibid.). Such attitudes can also be taken to reflect the view of personnel in the wider music industries, as they have "long been suspicious of academic courses and tend to value hands-on experience above qualifications" (Cloonan 2005: 86).

Though the value and perception of formal education programmes for road crew members are still being debated, they may be an asset for women interested in working in live music. The presence of women in sound engineering, and other technical roles, may be assisted by their ability to attend colleges and universities to acquire such skills. These institutions create opportunities for women to avoid the challenges, limitations and exclusionary practices that can occur with informal mentorship and on-the-job training.

More recently developed programmes offer hybrid versions of these courses. Bypassing the formal accreditation route, members of road crews have begun to offer free courses online in the form of Tour Mgmt 101.[8] This programme began in March 2020, shortly after the start of the COVID-19 pandemic, and is taught by industry personnel as a series of webinars about various aspects of touring. The timing of its origin can be seen as a way to generate or maintain interest in and connection to an industry that risked losing workers. Aspects of this programme will be discussed later in this book. The Tour Bus Smart programme[9] led by PiL/Pigface drummer Martin Atkins is a practical learning experience in affiliation with the Music Business Program at Millikin University in Illinois. Students do not have to be enrolled at Millikin to apply and can take it for college credit or not. Started in 2019, it places emphasis on extending teaching methods outside of the classroom through a "five-day immersion on a national concert tour." As evident by the website's description that "some things cannot be taught. Some things need to be *lived*..." the programme endeavours to cover the range of responsibilities, activities and emotions road crews confront on tour. This programme can be seen as a response to concerns or criticism over the disparity between classroom-based instruction and practical experience. However, how effective it is in doing so, beyond an introductory level, is difficult to gauge considering the potential limitations created by bringing a larger than normal group into the working environment of a tour. In this way, it may do more for the culture and experience aspects than in actual training. Similar to Tour Mgmt 101, Atkins also conducted free online sessions during the pandemic to teach students about touring, though he focused on grassroots-level musicians rather than on road crews. The next section of this chapter will discuss the manner in which road crew members initially acquire and continue to maintain touring work, or do not.

Getting In

Training and learning opportunities provide road crew members with access to a network, which is an essential component to acquiring work on tour. "Getting in" has been described as the most difficult aspect of working in the cultural industries (Blair 2001). Such a characterisation can also be said of the live music industry, given that the "concert business society has always been somewhat closed by its own nature" (Waddell, Barnet and Berry 2007: 2). Research interviews indicated that members of road crews hold a similar perspective on the industry.

> The touring business it runs in certain ways. Access to it is if you like by invitation only. You can't apply for a job and get it without somebody knowing you. I don't know whether that should change cause it works. Breaking into it's very difficult if you're not living in the right place, if you're not willing to work for nothing for a long period of time, you know.
>
> *(Joe 2019)*

Following their early work experiences, respondents identified a key opportunity, or a "break," that was integral to working on tour and that they perceive as the entry point to the career that followed. In all cases, respondents acquired their respective positions from someone they knew. Amy (2018) described how these networks start to develop.

> … you meet the same people and if you work in those jobs for a couple years, the same people come back, you build relationships, you're also building relationships with other people you might be working with eventually and the people you're working with locally. A lot of times they want to do the same things, so you're building future touring stuff.

While working as a production runner, she obtained a role as assistant TM through a band member who had performed at her venue. He put her in touch with his manager who was taking over tour management responsibilities and needed an assistant. As another example, monitor engineer Craig (2017) had been working as a drum technician for local bands when a friend of his, who worked for a local audio company,[10] broke his foot and needed someone to take over his job. And, while working at an event, Adrian (2018) received a job offer when someone he knew was notified that a crew member had to cancel. Because that person already knew Adrian and the quality of his work, he asked immediately if Adrian was available to start the following week. Claire Murphy (2019: 13) got her first job as a guitar technician while working in a guitar shop in London. A customer, who was a road crew member, told her that the band he was touring with needed a guitar tech at the last minute and asked if she would like the job. Despite not having a clear understanding of what a guitar technician was or did at the time, Murphy accepted. Tana Douglas (2021: 42–44) volunteered to load out a band's equipment one night at a local venue. When they returned to the area, they were short a crew member and asked her to help. Following this, they hired her to be a backline technician, which was her first job on a road crew.

In this way, road crews are consistent with workers in other sectors of the cultural industries who rely on networks of people (Blair 2001, McRobbie 2002, Hesmondhalgh and Baker 2011, Curtin and Sanson 2017). Members of road crews make their way in the field through a series of contacts made through and enabled by a variety of work experiences over time that helps them move into relatively steady freelance touring work. The importance of contacts to acquiring jobs emphasises the role of trust and functions as a form of "pre-screening" for people given the close-knit nature of working and living on tour, as the following chapters will show.

The transition from early work and learning opportunities to being a full-time freelance crew member is the product of a casual momentum rather than a clear-cut trajectory in which workers are aware of having completed or reached a particular stage. In this way, crew members often describe the sense that they "fell into" their careers rather than making a concerted effort to follow a particular path or attain a specific position. It suggests that careers on tour occur accidentally, were unplanned or came about easily and without forethought. The phrase "fall into" as a means to explain their career paths was noticeable in this study's sample and is commonly observable in interviews with or stories about road crew members in various media outlets. In general, however, members of road crews do not actually "fall into" their jobs, but rather are well positioned to acquire work in this sector based on their involvement in live music, their network of contacts and their pursuit of opportunities that then translate into steady freelance work.[11]

This notion, however, does seem to differ along gendered lines. While male road crew members commonly describe their experience of gaining access to the live music industry as an accidental chance of "falling into it," the language that women utilised represented a noticeable departure. Amy (2018) did not use this term to describe her path and presented a much more decisive and intentional approach to her early involvement in music. She stated that when she was working as a production runner and learning different jobs, she "wanted to tour" and made that known to PMs on incoming road crews. Tasha (2018) similarly did not use this term to characterise her experience. Claire Murphy (2019) also described being decisive, wanting to tour and repeatedly emphasised the importance of making what you want known to people in the live music industry and touring world. In her book, *Girl on the Road: How to Break into Touring from a Female Perspective*, she offers a step-by-step guide and systematic approach to getting into touring and managing life once becoming a crew member. While she states that the book is for anyone interested in a career in touring, that it is written in such a way and identified as being from a "female perspective" speaks to a particular experience and the factors accordingly deemed important. The difficulty for women in accessing the live music industry is almost made manifest in a more direct and purposeful approach. In its difference from the casual perspective described by men, their non-use of the term reflects the unlikeliness of a woman simply *falling into* this career path.

Getting Paid

Freelance work is often characterised by its relative precarity. The COVID-19 pandemic clearly illustrated that working in live music is marked by uncertain conditions and a range of challenges.[12] In recent years, discussions of the "gig economy" have pointed to a labour market distinguished by the "prevalence of short-term contracts or freelance work as opposed to permanent jobs" (OED, Cloonan and Williamson 2017) in which workers are subject to precarious and exploitative employment conditions. Uber drivers and food delivery services are common examples of work in the gig economy. Scholars have observed how a term rooted in popular music culture has been generalised to refer to wider societal and economic trends, but that, despite the relatively recent application, the "economy of gigs" has long characterised musicians' working lives (ibid.). My findings indeed establish that the working lives of members of road crews, like many types of cultural workers, do involve holding multiple jobs and self-employment, and that they encounter irregular work patterns and short-term contracts (Towse 1992, Hesmondhalgh 2013: 254). Working on a road crew also means contending with job insecurity and uncertain career prospects.

Early in their careers, crew members often work for low or no pay. As compensation, their food and travel are covered, and they gain experience. Many crew members can justify this aspect of their careers, see it as necessary and accept it as the reality of the live music industry. Murphy (2019: 17–18) described the compensation for one of her first tours as being unpaid, with food and travel covered, but that the work experience she received was "invaluable." Due to this, she is "not against working for free in the beginning" on the basis that experience is a necessary component to a future career and that it is very difficult to get paid while still learning on the job unless someone is "very lucky." Furthermore, she argues that money is only one form of currency and sees time as another. "If someone is allowing you to work with them and show you the ropes or give you knowledge, that is something they are giving you for free. Likewise, you are gaining in another way, just not monetary" (38). In Murphy's case, she was still learning her role and was shadowing another crew member, but this perspective in consistent with other workers in the creative industries regarding pay structures and expectations early in their careers.

This perspective on the value of working for free is not shared by all crew members. Tasha (2018) stated that throughout her career, she has always been compensated and has never worked for free. When she initially started, she was paid around $100 per day, a fee she views as "terrible money." Tasha advises new entrants to the field that they "should never work for free." Her advice is based on the risk that agreeing to low or no pay poses in setting a "bad precedent" that can have broader effects on workers by lowering the standards and expectations around acceptable pay. In the mid-1970s, Tana Douglas (2021: 50) earned $60 (Australian) per week as backline crew. The rate was for food and incidentals, with all other expenses being covered by artist management. Douglas characterised the pay as the equivalent of winning the lottery, but this was enhanced by the fact that she had no rent to pay and, as someone in her late teens, was new to the workforce and paid labour. In these ways, road crew members have differing attitudes about pay rates and their suitability relative to their own standards and concerns.

The topic of money was not addressed explicitly during interviews for this study as such questions can be perceived as intrusive and could have affected the willingness of respondents to participate. Reynolds (2012: 145) stated that there is "no international standard pay scale for touring road crew." Following this, estimated pay figures for road crew members are available, but they vary drastically. Figure 2.1 provides a sample of some of the average pay rates according to the size/scale of a tour, experience and specific roles. Rates are indicated as daily, weekly and annual based on reporting from available sources.

In contrast to the precarity that is so often associated, freelance work is also marked by relative stability that is the product of accumulated experience and forms of capital that are generated through reputation and networks of contacts. As Becker (1982: 86) notes, while workers may be judged on the basis of their last job (see Faulkner 1971) and are not protected from the effects of a poor work performance, some "manage to work regularly, moving from one project to another." The research sample demonstrates that self-employed road crew members can and do work consistently in the live music industry for decades and establish long-term work relationships with musicians. Short-term labour can therefore be the site of relative privilege and a setting in which workers rise to the top of their field (see Faulkner 1971: 45, Peterson and White 1979).

Freelance work can also enable access to a relatively comfortable lifestyle. Comments during interviews and participant observation, and supporting archival documents in the David Russell Collection,[13] gave a strong impression that members of road crews can be and are well paid professionals. Passing remarks were made about the locations of homes and the comforts and privileges their line of work has provided, remarks which support this perception. Crew members also made reference to learning to navigate uncertainty over

	Daily Rate	Weekly Rate	Annual
Entry Level – General	Unremunerated–£50/$75		
Small/Van Tours – General	Unremunerated–£125/$150	$700–$1,000 for small/club tour	
Bus Tours – General	£200–£300/$250–$350	$3,000 for international tour	
Established/Experienced with Specific Role	£150–£250+/$150–$350+		
Tour Manager		$2,500–$10,000	
Production Manager		$1,500–$8,000	
Road Manager			$25,000–$125,000
Front of House Engineer			$60,000–$120,000
Monitor Engineer			$35,000–60,000
Lighting Director		$500–$1,000	
Merchandise		$1,000	

FIGURE 2.1 Pay scale information.

Sources: Reynolds (2012), Berklee College of Music (2016), and Murphy (2019).

the course of a career with careful attention to money management. Such factors are often lost in discussions of freelance work but are important considerations in studies of cultural workers. The chapter now moves into an examination of the ways in which road crew members continue to acquire work and maintain a career.

Maintaining Work

This chapter has shown that gaining access to the industry is dependent on contacts and that this method is the preferred and trusted manner of doing so for members of road crews. As freelance workers, continuing to get jobs follows a similar informal method that is based on contacts and word of mouth. While musicians may hire crew members early in their careers, as they become more successful, they are less involved in doing so. The artist's management or the TM is responsible for hiring the crew (Reynolds 2008: 387). Crew members become aware of available positions not by job advertisements, which are rarely publicly posted, but through their network of contacts.

> Well, not one of these jobs would have been advertised anywhere. It's just 'I need a tour manager does anybody know anybody?' Then they'd be like 'oh well ask him,' 'oh I don't know, but I think he's available.'

> There's constant chat on Facebook, anybody got any work? I'm free. You know, I'm finished. Anybody know of anything? And that's, I got [job with artist] cause their road manager he said 'I can't do it but I know this guy is free'.
>
> *(Joe 2017)*

As this quote indicates, crew members are offered available positions based on recommendation and/or prior contact with the relevant parties. This process has been compared to the manner in which electricians or plumbers are hired. The most trusted way to find a plumber or electrician is to call someone who has recently used their services and ask for a recommendation. As example, Claire Murphy (2019: 18) explained that she was hired "purely" on the recommendation of a mentor/colleague. When hiring road crew members, enthusiasm and experience are regarded as less important than a verbal positive performance review (Reynolds 2008: 388). In other words, crew members are "only as good as your last call" (Faulkner 1971: 111). Craig (2017) summarised how this process typically works.

> Normally you just get a call, or these days more often an email, saying are you available, such and such, either from someone you know or someone saying such and such has recommended you. 'We need a guy for this touring period, are you interested,' and that starts the conversation and then assuming if all things are right, money's right, all the rest of it, then it either continues or it doesn't. But yeah, it's very direct and very informal.

Adrian (2018) confirmed a similar experience by explaining that "quite often you get calls from friends, 'are you available?'" If the crew member of interest is not available, that person will usually suggest several other people who may be. It keeps "the circle tight" (Douglas 2021: 68) and reflects the closed nature of road crews. As tours are scheduled and crew members accept work, finding acceptable and available workers can be difficult. Joe (2018) recalled asking a dozen US-based PMs to fill in for a five-week North American tour and learning that all of them already had work commitments.

The practices around the use of CVs further confirm the significance of the system of referrals in recruiting road crew members. My research indicates an inconsistent practice with the use of CVs. In general, just as jobs are not advertised, submitting CVs and formal applications are not the typical method of recruitment and hiring. Duncan (2018) described his experience:

> … its word of mouth or people you know rather than anything else, you know I've never had to give a CV to anybody. I believe people have started asking for them nowadays but I've never needed one, it was always

> somebody I knew would say 'oh this band needs somebody for this tour do you want to do it' and I'd say yes.

In some cases, a crew member will be asked to send a CV after already being referred for a job in order to demonstrate their experience.

> … if you've actually worked with somebody … somebody's been referred for a job and then, and honestly, typically in this business a resume is almost a formality, you get referred and somebody'll call you so you know they're interested in you and then they'll ask for your resume cause they wanna see what you've done, so that kind of speaks to the level of work that you can do …
>
> *(Amy 2018)*

The CV functions in such cases as a legitimating and supporting component to the recommendation rather than being the initial space of evaluation. Documents in the David Russell Collection archive exhibit further how CVs and cover letters function in the recruitment of crew members. Ten relevant documents were in the archive and were related to tours for Genesis and R.E.M. (Various CVs, David Russell Collection). The documents were usually sent to the artists' management, though one was sent directly to Russell. The context and circumstances of these submissions are not entirely clear, nor is the manner in which applicants were made aware of the positions. However, the wording of the letters, and the backgrounds of the applicants, strongly suggests they became aware from word of mouth, mutual contacts and as relative industry insiders. Two CVs appeared to be "cold calls" to inquire about the possibility of being hired and make the management aware of their skills and availability. One applicant seemed to have an existing connection to Russell, who expressed preference for her over another candidate, with whom Russell also seemed to be familiar. In three cases, the crew member submitting a CV and cover letter identified mutual contacts who served as a referral or had directly suggested sending an application. In these cases, details about the mutual contacts, and the circumstances and context of their working relationships, were emphasised and foregrounded in the written communication. For those without direct contacts, they attempted to overcome that absence with a variety of tactics, which revolved around some type of indirect reference or interpersonal communication. One applicant contacted the artist's management office prior to sending documents while another included a reference letter from the TM of the artist for whom he currently worked. In the latter case, his documentation was forwarded to Russell (via the management), at which point the applicant included a note that mentioned a mutual contact from whom Russell could obtain additional information about the applicant.

Other submissions clearly identified a key role or organisation within the live music industry with which they were affiliated and had been employed. Such an approach foregrounds and relies on the reputation of an organisation in order to compensate for a direct connection. These applications also featured a much more formal tone than those with mutual contacts. As these examples are solely connected to the hiring practices for crew members working for top-tier artists, they may also reflect an interest or need to recruit workers with a particular profile or may function as an additional screening process.

These findings suggest that referrals and mutual contacts form the basis for, and are the most important aspect of, the recruitment and hiring practices of road crews. Whether or not CVs and application documentation are needed or requested, the presence of a mutual contact is valued above all other considerations. This is made additionally obvious when applicants rely on other methods to generate a sense of legitimacy when contacts are not available. Similar patterns in recruitment practices have been identified amongst film production crews, which are almost entirely based on referral, formal HR procedures and CVs are rare or only provided after personal recommendation and job advertising is uncommon (Blair 2001). The practices in live music and film, however, represent a difference to practices in other cultural sectors. The touring productions of Cirque du Soleil, for example, rely on an HR department to recruit its employees.

The emphasis on networking and word of mouth for acquiring work is further evident in the connection between touring and job offers.[14] As with other types of freelance workers, road crew members compete for jobs in a labour market where "ability, reputation, tact, and social contacts determine the nature and volume of [their] work" (Faulkner 1971: 44). It can be said that touring is the manner in which members of road crews earn a living at the same time that it functions as the basis for how they continue to acquire jobs.

> It's all word of mouth ... it's all about relationships, it's all about connections, it's about people like seeing you do your job ... there's always this joke about take a short gig if you want work because for some reason if people see you working they ask you, but it's all about keeping your relationships and people, it's really like when you're working around 20 other people on tour, and you're working with different management companies and ... sometimes there'll be management companies or agents that ... have another band that they just happen to know from the management that they're looking for a good tour manager.
>
> *(Amy 2018)*

As Amy's quote indicates, when crew members are on the road, they are actively maintaining and forging relationships with people who are witnessing and evaluating them do their jobs. Networking is a practice that members of road crews understand as simply being built into the nature of their jobs more than the kind of deliberate act it can be for other types of workers.

> Work came to find me. And it all came from networking and I didn't know I was networking. At the time I probably rejected the term, it was far too business-like. Networking, don't be ridiculous. But I was out on the road all the time, meeting people, talking to people, looking useful, being useful, saying yes, doing favours ... You remember those kinds of things ... Cause you never know that might be a new job for you soon.
>
> *(Joe 2017)*

> Generally, once you're in, you're in. Of course one band alone isn't going to keep you employed for your entire career because all bands need downtime ... [Her mentor] continued to recommend me for shows he couldn't do, so I started getting experience with all types of different bands and making new contacts ... Not long after that I was recommended for a new tour by a colleague who used to rent space in my warehouse (it's all about connections and making sure those connections know what you want or have to offer).
>
> *(Murphy 2019: 20, 24)*

Rather than engage in workplace interaction as a "guise [for] networking" (Hesmondhalgh and Baker 2011: 155), the latter is an informal product of the proximity and collegiality that characterise road crews and working on tour. At the same time, nurturing and maintaining those relations is integral to acquiring future work, as it is that network upon which crew members depend. Helen Blair (2001: 160) observed a similar pattern amongst film crews, in which people known from previous working relationships or via references are hired, and those who are not known are excluded. After crew members gain access, former colleagues become the primary type of contact and replace previous methods of securing work, though friends can remain important sources of referrals and work-related information (ibid.). Road crew members engage in a continuous and mutual process of evaluation that is a factor in how they continue to acquire work. As Chapter 4 will show, this built-in system of networking and evaluation can and does have implications for how road crew members conduct themselves and experience being on tour.

My findings indicate a direct connection between a crew member's presence and visibility on what they call the "touring circuit" and the receipt of job offers and ability to maintain steady work. Crew members are aware of each other's active presence on tour through word of mouth, social media, at music festivals and through association with the artists they work for. Being on tour means "you're out" and that "people know you're out" which also implies that a crew member is actively looking for work. Similar to film crews, workers hear of and secure work through personal contacts who provide recommendations (Blair 2001: 152). If mutual contacts know of other jobs that are or will be available on upcoming tours, the crew members already on the road become the most likely candidates for those positions.

This aspect of their working lives reveals the risk of job insecurity and uncertainty when taking time off from touring. Aside from normal breaks and holidays, if they are or must be away from the road, it becomes more difficult for road crew members to access work again when they are ready to return. Being off the road means crew members are "off the game" or "not part of the scene anymore" and they "just disappear" (Craig 2017). Tana Douglas (2021: 47) recalled being advised early in her career that "It's always better to go from one band to another without any downtime. Downtime can kill you." Part of why this occurs is because the "relations among people in crews are typically defined by their function in the crew with the assumption that one member can be easily replaced by another member with the same skill set" (Arrow and McGrath 1995: 389, Webber and Klimoski 2004: 266). It is further enhanced by the oversupply that is a characteristic of the cultural labour market, which means that workers can quickly be replaced. Craig (2017) stopped touring in 2013 in order to spend more time at home and work on a major event being held in his local area. He did not intend to leave permanently but continued to receive steady work in his local area and ultimately did not go back on tour. Reflecting upon when he initially left, he stated:

> It's amazing how quickly when I stopped touring, how quickly the phone stopped ringing and the emails stopped. Within under a year, nine months, and I haven't had an offer. I think I had one offer last year.
>
> *(Craig 2017)*

Such working conditions are particularly problematic when crew members become sick, injured or need to take time off for personal reasons. Tony (2018) explained how he was affected after taking time off following the birth of his daughter.

> I was in a very steady employment for an artist when my daughter was born, she's now 9, but when she was born, I decided to take three months off to obviously help my wife and spend some time on the farm and that sort of stuff. It actually stretched on a little bit more, four months, but at the end of that four months I found it incredibly hard to get anybody, cause once you make a gap that gap gets filled by somebody else and it's one of

> these situations where there's a lot of social aspects of it, so if that person comes in and fills your space, makes friends, you get to a point where it doesn't matter how good you were, this guy's now here, so sorry that's just the way it is. So no, there's no job protection at all.
>
> *(Tony 2018)*

Joe (2017) was unable to tour for a period of time following an injury incurred from an accident. His experience of coming off the road was similar to Craig's.

> Put it this way, I'm not getting any job offers anymore, but I'm not on the road anymore. At least I'm not getting many. Cause people know I've come off the road, I think, and that's not necessarily been on social media.
>
> *(Joe 2017)*

In contrast, he did return to touring when he was offered a position as a PM, which he attributed to being "very lucky." He described his experience of returning to the touring circuit.

> … because now I've been back out on the road, since June, job offers have just started coming in again. I've been offered three or four decent gigs, that I have not been able to do because I'm doing this. And once those start bouncing around and there's a little network of people who start talking to each other, they go 'oh you're out, you're looking' and there's three weeks in the middle I can't do so I've got people doing them for me.
>
> *(Joe 2018)*

Visibility on the touring circuit essentially translates into access to work hours and varying degrees of job uncertainty. This suggests that crew members' networks are complex and ephemeral given that the networks are most effective when crew members are actively working on tour and leaving the road represents a break. The nature or strength of relations between workers are less important and influential than are presence and visibility. Road crew members on tour comprise the "core workforce" that has the greatest access to work hours and less employment uncertainty due to a network of personal connections. Those that leave the road temporarily become the "peripheral workforce" that encounters greater uncertainty and the potential for longer periods without work given that they have lower access to work on the basis of limited contacts (Christopherson and Storper 1989, also Blair 2001). Road crews can and do move between these two categories. These terms are adapted and applied here to the live music industry from Susan Christopherson and Michael Storper's study of film crews. The concepts of the core workforce and peripheral workforce are useful for illustrating this aspect of road crews' working lives. However, the authors equate "experience" for film crews with union accreditation that does not apply to road crews. No official union for members of road crews currently exists, though there was an attempt in the United Kingdom in 2003–2004 with the short-lived Roadcrew Provident Syndicate. My research suggests that the difficulty of organising is linked to the freelance, self-employed status and independent, self-sufficient mindset of road crew members. Road crew members can join the International Alliance of Theatrical Stage Employees[15] in the United States and Canada, or the Broadcasting, Entertainment, Communications and Theatre Union[16] in the United Kingdom, of which only one respondent reported being a member. They can also join and consult the Production Services Association,[17] which is a UK-based trade association organised to address the needs of the live event production industry and represent and advocate for workers.

The network that members of road crews depend on to secure work is also their competition. This is most clearly expressed when a crew member has conflicting commitments or is otherwise unable to take a regular job. In this situation, crew members ask people they know who work at the same level to temporarily cover for them. This situation comes with a risk of being replaced permanently, which can and does happen.

> Somebody you've called and said 'can you cover for me' has gone in and loudly criticised everything you've done and loudly said how much they could do it better and they ended up with the job. So that's dangerous … Yeah, I have people who are my competition, who were my contemporaries and now I don't necessarily have much contact with. Who have stolen work from me, I would regard it as, taken work off me. Some of them are doing well, some of them aren't. We're civil to each other when we see each other.
>
> *(Joe 2018)*

When temporarily filling in for someone is successfully handled, it works because crew members adhere to particular expectations. Crew members state that there is essentially nothing they can do to protect their jobs or prevent someone from taking their position. Rather, they become selective with respect to whom they entrust with such tasks.

> You just develop … a relationship with people you trust won't do it. Who will rather not do it, who will come in and do what's necessary, not rock the boat, not bad mouth you to the artist and get on with it. Because you need those people, they need you, you know. I've got three or four people in mind who've covered for me and I've covered for them and that works beautifully.
>
> *(Joe 2018)*

The arrangement also depends on a code of conduct and an understanding between crew members. It involves accepting, respecting and trusting the working practices of the person whose role is being fulfilled and not questioning their choices in front of the artist.

> And I've gone in and thought 'this is a bit odd' I wouldn't have done it like this but I keep my mouth shut, do the four weeks, get off again, maybe chat to them about it, 'why'd you do it?' there might be a reason for it. Maybe an artist-driven reason why things are done this way and not another. So, you just trust them and get on with it.
>
> *(Joe 2018)*

Claire Murphy (2019: 18) was confronted with a similar situation early in her career. After being offered the chance to shadow a guitar technician, another colleague asked her if she was going to stay on the tour after her mentor had left. She saw this practice as unfair as it would involve taking her mentor's job. By adhering to these practices, crew members create, rely on and maintain informal systems that help them navigate their career paths while they also protect each other's reputations and livelihoods.

Research respondents in this study spend an average of six to nine months working each year, and their trajectories reflect the working patterns of a group of crew members with decades of experience. They are typically hired as independent contractors for a specified period of time. Respondents noted that they generally do not receive work contracts when being hired for a tour. The work agreement is made via an email exchange in which the terms are indicated, such as the role, pay and amount of time. Practices can, however, vary. Respondents also reported that it is common to be put on payroll for the duration of their hire when touring with American artists. They characterised the early years of being freelance workers as difficult and stressful, since starting out working for smaller bands with limited tours made it hard to put together a schedule and a sufficient income. In other words, it took time, experience and contacts to accumulate enough work and this type of schedule. The distribution of work over the course of a year is dependent on several factors. The first is that work logically follows the relative high and low seasons of the live music industry. Live music seasons are marked by "practical weather conditions for leisure arrangements. Outdoor music festivals are concentrated in the summer season, bracketed by the club concert touring seasons from March to May and September to November" (Holt 2010: 245). This factor was reflected in research findings as respondents addressed how their schedules are shaped by the time of year in relation to peak times for tours. Adrian (2018) described a "typical year … would be February to May, then maybe a short break, very very busy summers with the amount of festivals that there is now in Europe … and September to December is very very busy again." Another factor is that road crew members tend to work for several different artists and ultimately depend on this variety to make a living.[18]

> ... it tends to be the same four to six artists recurring, cause once they're on an album cycle you can do anything from four months to a year with one and by the time you've done that with a couple of artists the first ones have come round again. So certainly over the past ten years or so it's been the same half a dozen artists recurring.
>
> *(Adrian 2018)*

This is directly related to how musicians work and earn a living as their careers are marked by, and involve balancing, a "particularly relentless routine of touring and recording" (Frith et al. 2019: 166). Ideally, one artist is recording while another is touring, which facilitates continuity in crew members' work schedules. An overlap between artists' schedules puts crew members at risk for losing work or compromising their working relationships when having to choose between tours.

> ... the nature of bands is they don't tour constantly, they tour for a year and then go in the studio for a year or two. So if you're lucky your bands match up, so I worked for [artist] for 17 years, I worked for [artist] for 10 years. It just so happened that their touring schedules didn't overlap so I was able to keep those going. It's when they overlap that you either have to give it up or get someone else in to cover you otherwise you might lose that band or whatever.
>
> *(Craig 2017)*

Crew members' touring schedules are ultimately shaped by and dependent on the artists for whom they work. In this way, they generally have little or no discretion or control over their own schedules. In one case, however, a respondent explained that she carefully plans her schedule in order to accommodate other activities. Tasha (2018) stated "ideally I try not to work more than six months out of the year ... I would just turn down work from November till probably March every year" which allowed time to pursue other interests. Tasha balanced this arrangement by working during the busiest months for tours which in turn facilitated the ability to take time off. The next section of this chapter will explore the topic of getting fired from a road crew and the factors that enable or prohibit such action.

Getting Fired

Consistent with how contacts are integral to acquiring work, research findings show that interpersonal relationships factor significantly into a decision to fire a member of a road crew. Getting fired or retaining a position is directly associated with the strength of their relationships with an artist. In particular, the likelihood of getting fired is greatly reduced if the road crew member in question is viewed favourably by an artist. During participant observation, a local production manager explained that a crew member working on that day's concert had a drinking problem and that the other crew members had tried to get him fired but were unable to do so. The local sound engineer, who also freelances on tours, described retaining a position because a group's lead singer preferred her though the keyboard player preferred a different person. Both she and the local production manager agreed that "there's a lot of that kind of stuff that goes on." Joe (2018) further confirmed these types of practices.

> ... you often find it very difficult if not impossible to fire someone that [the artist] like ... Yeah, unless you've got a very good reason for it, and then they should trust you, but they might still overrule you, 'yeah, he was drunk last night, but I love having him on the road, he makes me feel good, keep him' ... Yeah, cause I don't think he's doing his job properly. And often the face they show the artist is fantastic and then they're not loading the truck, or they're not doing what they need to do, you know, communicating properly. So at that point then it becomes very difficult. But an artist who trusts you will go 'well if that's what you really think, then he's got to go.'

The working relationship and clear communication between the TM/PM and the artist are integral in attending to firing practices. However, a complicating factor is that the crew member in question may behave differently towards the artist than towards the rest of the crew. Such impression management conceals the reality of the crew member's job performance and gives the artist a favourable perception. It is in such circumstances that artists must rely on the trust they accord to TMs and PMs in attending to performance evaluation and making decisions.

Research respondents also described contrasting circumstances, however, in firing practices and artists. Tasha (2018) prefers not to work closely with artists due to how it makes her feel as though she has to be on-guard and places her "in the firing line." This perspective comes as a result of having witnessed a variety of "people get sacked for no reason because an artist is just temperamental." She recounted one situation in particular.

> ... I had a set carp[enter] who walked past one of the artists in an arena, and I guess the guy said hi to the carp and the carp didn't respond cause he was probably in the middle of something, you know, was thinking about fixing some problem, whatever, didn't acknowledge the artist and got fired

> because of it. It's like when you're around the artist you're just that much closer to being fired for any irrational reason.

Tasha's preference to avoid working with artists is not experienced in isolation. The aforementioned Tour Mgmt 101 features a document on its website titled "Rules of the Road"[19] that provides guidance on crew conduct. It offers insights that resonate with Tasha's experience and concerns over the likelihood of getting fired when working closely with musicians.

> The stars are not our friends they are our employers. 'Closest to the fire, first to get burned.' If they know your name, guess who they're going to freak out on when something goes wrong? Also, it wouldn't be a bad idea to remember exactly who you work for.

Andy (2017) similarly described the ways that artists utilise the power to fire on members of road crews, which he identified as being connected to particular emotional states. The artist's excessive use of firing resulted in the crew being unable to take it seriously and reappropriating it as humour. Such practices suggest a disregard for the crew's roles, a belittling of their job security and reinforce the live music hierarchy.

> There's artists who every single performance there's self-doubt, almost loathing, then you mix in drink and drugs and you just get fireworks, then it's like how do you deal with that, everybody gets fired, there was one band I toured with and it was a standard joke we'd all get fired every single night and then next morning the artist would come out and say 'you know I'm so sorry' and all this. And you know it's just ... at that level it's a bit of a joke, but then it can really go really badly.

In these cases, artists' behaviour represented a clear hierarchical division and a subordinate view of crew members by utilising power out of self-interest and to fulfil ego-driven needs. That instances such as these occur and negatively impact the working lives of road crew members also shows the extent of the effects of working on tour. Musicians encounter and experience difficult working conditions, pressures and insecurities that come with the potential to be misdirected and further exacerbated with substances (see Wills and Cooper 1988, Toynbee 2000: xi, Jones 2012: 70). Despite musicians occupying positions as employers, these situations also represent a problematic understanding of and inexperience with business practices in the music industries (Jones 2012: 70; see Chapter 4).

The usual practice for firing is that the crew member in question will finish the tour and not be asked to return for the next one. A crew member is effectively fired when they are not called back or are told that the arrangement was not working, and a decision has been made to select someone else. Firing someone during a tour presents logistical challenges for TMs.

> Actually on the road said you're going home? ... Truck driver who was drunk. Marched him to the train station, said cheerio ... three or four times, to be honest. That's it, you've got to go and I've got to find somebody for the next show. And that's often the defining moment. I mean South America everyone's got a visa, there's absolutely no way of replacing this person without a massive headache, it's just not gonna go well.
>
> *(Joe 2018)*

The itinerary of a tour makes it difficult to fire a crew member abruptly due to the fact that they need to be quickly replaced. This is particularly difficult in countries where crew members depend on visas to be able to work. However, in extreme circumstances, these measures are necessary for the wellbeing of the touring party and safe continuity of the tour. Practices around firing reinforce the nature of crews, as work groups, being able to sustain regular changes. At the same time, the challenges in firing practices illuminate the influential role of a tour's itinerary and the logistics of travelling in the inability to replace crew members when needed.

Gender can also factor into being fired from a job. This chapter discussed various reasons that contribute to why road crew members lose their positions. While an artist's preference for particular crew members is among them, this can be complicated by gender. Tasha (2018) was once fired from an otherwise all-male crew and artist and replaced by a man because she is a woman. The exact reason for the firing was not made explicit, but gender was certainly the determining factor.

> ... for whatever reason I was stuck on the band bus and it was an all-male band and it just kind of got around, I don't know if it was girlfriends and wives or whatever, they just didn't want a woman on their bus and there wasn't room for me on the crew bus and I was replaced by a man ... It was explained to me by the tour manager. Which was nice because sometimes you just don't even get that explanation, you're just gone. But it was really kind of him to explain 'oh it's not you, it's really not you, it's strictly like a weird gender thing and it's like fickle artist's garbage.'

In this case, relative discomfort with their gender renders a crew member unfit for a job. The decision to fire the crew member is not based on whether or not the artist likes or dislikes her, but due to an aspect of her identity that is deemed incompatible. John Vilanova and Kyle Cassidy's (2019: 96–97) study on "merch girls" included a similar account of losing work on the basis of gender and that women's presence on tour is unwanted because they are "presumptuously sexualized." This instance further illuminates the kind of sexism women can encounter and the difficulties of being a minority on tour.

Leaving the Road

Leaving the road is also a factor in the working lives of road crew members and one that respondents confronted and addressed. Research findings indicate that leaving the road can be a product of extenuating circumstances that are unrelated to the everyday working conditions and realities of touring. Two crew members stopped touring for health-related reasons; Amy (2018) quit when she developed a health problem and was forced to stop touring, and Joe (2017) left after sustaining an injury. The former has not returned to touring, though she still works in the industry in non-touring roles, and the latter returned after recovering. Joe tried other occupations, such as teaching, but "wasn't happy. I like being on the road and I'm good at it. And I've tried a few other … things and I'm less good at it and less happy." He stated that as he is approaching 50, he has thought about retirement but believes he will continue as long as he still enjoys touring. As previously mentioned, Craig (2017) stopped touring in 2013 to work on a local production and spends more time at home after purchasing a new house with his wife. While he was touring, he never considered changing careers. He "didn't actually intend to stop permanently, I intended to stop for a year and then see what happened" but ultimately did not return and acquired a position with his wife's live events company. Though Craig's reason for leaving was influenced by the amount of time he was away from home, such conditions were ultimately not the determining factor nor were they seen as a reason to quit permanently. At the time of interviews, Tony (2018), in his late 40s, planned to leave after the artist for whom he has worked for the past 17 years finishes the next tour. He noted that he expects it to be a "hard transition." These cases demonstrate that crew members enjoy the type of work they do, continue if they can or find work in a related area.

Road crew members often struggle with whether or not to remain in their line of work. Questioning the norms and working conditions of a career as a crew member and life on the road is a consistent pattern. Hawes (2019: 57) encountered what she referred to as the "final straw" and nearly quit after an event organiser was only willing to speak with her after 4:00 a.m. She was deterred from quitting, however, when the person she'd been working for questioned which other type of work would pay so well, provide the same lifestyle and offer similar excitement (ibid.). Tasha (2018) stated that she's "probably tried to quit working on the road five or six times" and would like to stop touring full time. Part of the reason is directly related to the working conditions of touring, which she described as "physically taxing, the travel is crazy, the lack of sleep is crazy. I've had heath issues I may not have had because if I hadn't worked on the road and yeah, I'm just tired." It is also connected to her age (mid-40s) and the labour oversupply, as younger people can be hired in her position for less money. Tasha would prefer to "keep my foot in the door" while doing other types of work and described coming up with an "escape plan" that would allow her to stay partially connected to touring and road crews as an accountant while being able to spend more time at home. Her situation provides insight into the difficulties that members of road crews confront as they get older and become aware of the realities and potential limitations of working on the road with possibilities for moving into new positions outside of touring. Her use of the phrase "escape plan" characterises this difficulty by invoking a desire to leave and to locate the means to do so. The lifestyle of touring is a contributing component in other cases. Andy (2017) reduced his touring schedule when he "got to [my] forties and the attendant problems of drinking too much and drug abuse, obviously take a toll." He began writing books about working on tour and now teaches music business courses for most of the year and tours during the summer season. Both cases demonstrate that crew members take advantage of the skillsets they acquire from touring and transition them into new but related types of work. The chapter will now close with a summary and conclusion.

Conclusion

This chapter has provided an overview and analysis of how road crew members become part of live music industry. They acquire skills and become oriented to live music through interests and pursuits that create the conditions of possibility for their future careers. Working at venues functions as a legitimating factor in the choice of live music as a career, provides training and places workers within a network of contacts. There are no formal qualifications for becoming a member of a road crew, and the emergence of university training programmes is viewed with suspicion. Such a perspective shows that crew members accept and trust the informal process of learning that occurs on the job and through the accumulation of practical experience. Road crew members strongly rely on contacts and informal practices to gain and maintain work. They get in, move through and advance via a series of work opportunities and their network. Likewise, contacts also matter when crew members are confronted with an inability to work, or risk being fired. In these ways, the informal practices that mark the paths of crew members and the networks they form and rely on are the essential components to gaining and maintaining touring work. The next chapter will discuss and analyse the various components involved in a typical show day on tour.

Notes

1 See the podcast Roadie Free Radio: http://www.roadiefreeradio.com/
2 See also Gorman (1978: 38).
3 See Music Venue Trust: musicvenuetrust.com.
4 See Faulkner (1971: 113) for evidence of a similar pattern amongst freelance studio musicians.
5 https://academyoflivetechnology.co.uk/
6 https://www.acm.ac.uk/courses/higher-education/technical-services-route/
7 https://www.productionfutures.co.uk
8 https://www.tourmgmt101.com/
9 https://toursmartbus.com/
10 Some road crew members work for audio or lighting supply companies prior to working directly for artists, and some audio and lighting technicians work solely for such companies. An in-depth examination of such companies and career paths is outside the scope of this book.
11 See Webster (2011: 127–152) for an analysis of the ways in which concert promoters depend on networks and accumulate social capital to maintain their careers.
12 For an understanding of the variety of financial support available to freelance road crew members, and the attendant issues, see Barna (2020), Barna and Blaskó (2020) and Mufamadi and Koen (2021).
13 David Russell was a tour manager and production manager for more than 30 years. An extensive archive that contains documents from his career is available at the Rock and Roll Hall of Fame Library & Archives. More information on the David Russell Collection is available here: https://catalog.rockhall.com/rrhof-ais/Details/archive/110000031
14 This is different than the selection criteria for other types of crews, which tends to be "highly focused on task-related qualifications" (Webber and Klimoski 2004: 267).
15 https://www.iatse.net/
16 https://bectu.org.uk/
17 https://www.psa.org.uk/
18 See Faulkner (1971: 113) for evidence of a similar pattern amongst freelance studio musicians.
19 Available at https://www.tourmgmt.org/templates.html

References

Ames, Richard. 2019. *Live Music Production: Interviews with UK Pioneers*. New York: Routledge.

Arrow, Holly and Joseph E. McGrath. 1995. "Membership Dynamics in Groups at Work: A Theoretical Framework." In *Research in Organizational Behavior* 17(1): 373–411.

Barna, Emília. 2020. "IASPM Hungary: Developments and New Directions in Popular Music Research." *IASPM Journal* 10(1): DOI 10.5429/2079-3871(2020)v10i1.7en.

Barna, Emília and Ágnes Blaskó. 2020. "Music Industry Workers' Autonomy and (Un) Changing Relations of Dependency in the Wake of COVID-19 in Hungary: Conclusions of a Sociodrama Research Project." *Intersections. East European Journal of Society and Politics* 7(3): 279–298.

Bayton, Mavis. 1997. "Women and the Electric Guitar." In *Sexing the Groove: Popular Music and Gender*, ed. Sheila Whiteley, 37–49. London: Routledge.

Becker, Howard S. 1982. *Art Worlds*. Berkeley: University of California Press.
Bell, Thomas L., Brian Bell and Carl Taylor. 2022. "Impacts of the COVID-19 Pandemic on Live Musical Performance in the United States and the United Kingdom: Outsider and Insider Perspectives." In *COVID-19 and a World of Ad Hoc Geographies*, eds. Stanley D. Brunn and Donna Gilbreath, 1927–1953. New York: Springer.
Berklee College of Music. 2016. "Music Careers in Dollars and Cents." The Career Development Center. https://www.berklee.edu/sites/default/files/d7/bcm/Music%20Careers%20in%20Dollars%20and%20Cents%202016-rh.pdf.
Blair, Helen. 2001. "'You're Only as Good as Your Last Job.': The Labour Process and Labour Market in the British Film Industry." *Work, Employment & Society* 15(1): 149–169.
Bourdage, Monique. 2010. "'A Young Girls Dream': Examining the Barriers Facing Female Electric Guitarists." *IASPM Journal* 1(1): 1–16.
Christopherson, Susan and Michael Storper. 1989. "The Effects of Flexible Specialization on Industrial Politics and the Labor Market: The Motion Picture Industry." *Industrial and Labor Relations Review* 42(3): 331–347.
Cloonan, Martin. 2005. "What Is Popular Music Studies? Some Observations." *British Journal of Music Education* 22(1): 77–93.
Curtin, Michael and Kevin Sanson, eds. 2017. *Voices of Labor: Creativity, Craft, and Conflict in Global Hollywood*. Oakland: University of California Press.
Douglas, Tana. 2021. *Loud: A Life in Rock 'N' Roll by the World's First Female Roadie*. Sydney: ABC Books, HarperCollins Publishers.
Faulkner, Robert R. 1971. *Hollywood Studio Musicians: Their Work and Careers in the Recording Industry*. Chicago: Aldine.
Frith, Simon. 2007. "Live Music Matters." *Scottish Music Review* 1(1): 1–17.
Frith, Simon, Matt Brennan, Martin Cloonan and Emma Webster. 2013. *The History of Live Music in Britain, Volume I: 1950–1967*. Farnham: Ashgate.
Frith, Simon, Matt Brennan, Martin Cloonan and Emma Webster. 2019. *The History of Live Music in Britain, Volume II: 1968–1984*. London: Routledge.
Gorman, Clem. 1978. *Backstage Rock: Behind the Scene with the Bands*. London: Pan Books.
Gross, Sally Anne and George Musgrave. 2020. *Can Music Make You Sick? Measuring the Price of Musical Ambition*. London: University of Westminster Press.
Hawes, Kim. 2019. *Confessions of a Female Tour Manager*. Independently Published.
Hesmondhalgh, David. 2013. *The Cultural Industries*. 3rd ed. London: Sage.
Hesmondhalgh, David and Sarah Baker. 2011. *Creative Labour: Media Work in Three Cultural Industries*. New York: Routledge.
Holt, Fabian. 2010. "The Economy of Live Music in the Digital Age." *European Journal of Cultural Studies* 13(2): 243–261.
Horning, Susan Schmidt. 2004. "Engineering the Performance: Recording Engineers, Tacit Knowledge and the Art of Controlling Sound." *Social Studies of Science* 34(5): 703–731.
Jones, Michael. 2012. *The Music Industries: From Conception to Consumption*. Basingstoke: Palgrave MacMillan.
Kearney, Mary Celeste. 2017. *Gender and Rock*. Oxford: Oxford University Press.
Leonard, Marion. 2017 [2007]. *Gender in the Music Industry: Rock, Discourse and Girl Power*. London: Routledge.
McGinn, Matt. 2010. *Roadie: My Life on the Road with Coldplay*. London: Portico Books.

McRobbie, Angela. 2002. "Clubs to Companies: Notes on the Decline of Political Culture in Speeded Up Creative Worlds." *Cultural Studies* 16(4): 516–531.
Mufamadi, Kgomotso and Louis Koen. 2021. "Independent Contractors and Covid-19 Relief: Tax and Social Insurance Legislative Reform to Extend Protection to Independent Contractors." *South African Journal on Human Rights* 37(2): 277–301.
Murphy, Claire. 2019. *Girl on the Road: How to Break into Touring from a Female Perspective*. Independently Published.
Oxford English Dictionary (OED). 2020. s.v. "Gig Economy," accessed March 13, 2021, https://www.lexico.com/definition/gig_economy.
Peterson, Richard A. and Howard G. White. 1979. "The Simplex Located in Art Worlds." *Urban Life* 7(4): 411–439.
Reynolds, Andy. 2008. *The Tour Book: How to Get Your Music on the Road*. Boston, MA: Cengage Learning.
Reynolds, Andy. 2012. *Roadie, Inc.: How to Gain and Keep a Career in the Live Music Business*. 2nd ed. CreateSpace Independent Publishing Platform.
Smaill, Adele. 2005. "Challenging Gender Segregation in Music Technology: Findings and Recommendations for Music Education and Training Providers in the North-west." A report for the Regional Equality in Music Project, University of Salford.
Towse, Ruth. 1992. "The Labor Market for Artists." *Richerce Economiche* 46: 55–74.
Toynbee, Jason. 2000. *Making Popular Music: Musicians, Creativity and Institutions*. London: Arnold.
Various CVs, David Russell Collection, Library and Archives, Rock and Roll Hall of Fame and Museum.
Waddell, Ray D., Rich Barnet and Jake Berry. 2007. *This Business of Concert Promotion and Touring: Practical Guide to Creating, Selling, Organizing, and Staging Concerts*. New York: Billboard Books.
Webber, Sheila Simsarian and Richard J. Klimoski. 2004. "Crews: A Distinct Type of Work Team." *Journal of Business and Psychology* 1(3): 261–279.
Williamson, John and Martin Cloonan. 2016. *Players' Work Time: A History of the British Musicians' Union, 1893–2013*. Manchester: Manchester University Press.
Wills, Geoff and Cary L. Cooper. 1988. *Pressure Sensitive: Popular Musicians under Stress*. London: Sage Publications.
Wright, James "Tappy" and Rod Weinberg. 2009. *Rock Roadie: Backstage and Confidential with Hendrix, Elvis, The Animals, Tina Turner and an All-Star Cast*. London: JR Books.

Interviews

Adrian. Guitar Technician. In-Person Interview, 24 May 2018.
Amy. Tour Manager. Skype Interview, 9 July 2018.
Andy. Tour Manager and Sound Engineer. In-Person Interview, 18 April 2017.
Tasha. Crew Member. Phone Interview, 14 August 2018.
Craig. Monitor Engineer. In-Person Interview, 4 May 2017.
Duncan. Guitar Technician. Phone Interview, 10 May 2018.
Joe. Tour and Production Manager. In-Person Interview, 2 May 2017, 14 May 2018. Skype Interview, 28 August 2019.
Tony. Close-Protection Security, Head of Security. Skype Interviews, 4 August 2017, 16 March 2018.

3

SHOW DAYS

Working in live music is often perceived and mythologised as a glamorous and exciting endeavour. What is forgotten underneath such perspectives is that concerts, for those involved in the activities that lead to their realisation, are first and foremost work activities centred around a series of highly time-sensitive events. This chapter explores the daily activities that are integral to a show day on a concert tour and the working lives of road crew members. In doing so, it highlights the long-hours culture that is a key characteristic of their working lives, which is consistent with the cultural industries more generally. In this way, live music events are realised through the efforts of crew members that span lengthy working days. This chapter also describes the tour manager's and production manager's temporary workspaces and the significant activities occurring within them, which reveals much about the nature of their working lives and the culture of touring. Related to this are the important roles of various types of documents that help organise a show day and longer concert tour and function as essential communicative devices in the backstage area.

The Working Day

Individual concert events – or "show days" – are central sites of activity in the realisation of live music. Findings from participant observation indicate that it takes on average 12–14 hours on a show day to produce a 90-minute concert. Figure 3.1 provides an example. It focuses on the activities, as observed, that occur during a show day at a venue on a tour. It does not, however, account for the hours of work invested off-site, in hotels or on the bus, by various members of the touring party. Nor does it account for the hours worked prior

DOI: 10.4324/9781003303046-4

Example of Show Day

10:00 a.m. Arrival of Production Manager (PM) and technical crew members of touring party.

10:05 a.m. PM assesses the stage and designates where equipment is to be placed. Load-in begins; local crew assist with loading in equipment from the truck into the venue.

10:22 a.m. Riggers start working on stage backdrop and lights; this is handled by five people.

10:28 a.m. Set up of front of house soundboard begins; involved personnel consists of the band's Front of House (FOH) engineer and two local sound engineers. Merchandise person starts sorting through boxes.

10:31 a.m. Riggers finish attaching backdrop to rig. PM is briefed by local crew.

10:37 a.m. Backdrop rig is raised; stage lights are turned on.

10:40 a.m. Drum kit starts to be assembled on stage, which is placed on top of a riser.

10:43 a.m. FOH engineer starts to make adjustments on the soundboard; crew member instructs local crew on how and where to arrange and store empty flight cases in the venue until they are needed for load-out.

10:46 a.m. Props and on-stage lighting fixtures are placed on to the stage.

10:49 a.m. Small barricade is set up around the front-of-house area to separate it from the standing-room-only floor.

11:00 a.m. The house PA is turned on; amplifiers are placed on risers on the stage.

11:07 a.m. Barricade that separates the audience and the stage begins to be constructed.

11:14 a.m. On-stage lighting fixtures are tested; drum technician cleans and polishes the drum kit.

11:27 a.m. Recorded music is played over the PA to check sound levels.

11:30 a.m. The last flight case is removed from the stage and the ramp to move them is removed. Curtains have been hung at the wings of the stage; microphones begin to be set up by the crew.

11:35 a.m. Set up of audience barricade is complete; drum technician sets up glass enclosure around drums. Local crew place red rectangles on the stage side of the barricade. Bottles of water are placed by the barricade for the security to give to the audience as needed during the show.

11:40 a.m. The line check begins with the drum technician playing the drums.

11:43 a.m. The stage lights are turned on and checked.

FIGURE 3.1 Example of a show day schedule.

11:45 a.m. Crew members begin checking the artists' microphones.

11:52 a.m. Two crew members are on stage; one works on an amplifier, the other on the lead singer's microphone.

11:54 a.m. Guitar technician brings guitar on stage; additional microphone checks are done for each band member's set up, which involves communication between the instrument technicians and the FOH engineer. The drum technician checks the sound of all parts of the drum kit.

12:05 p.m. The FOH engineer encounters technical difficulties as the power suddenly goes out.

12:11 p.m. Power is restored.

12:13 p.m. Line check is complete.

12:15 p.m. Touring party leaves the area.

12:30 p.m.–2:00 p.m. House lights are turned off, stage lights are illuminated. Venue space is generally quiet except for checks and adjustments being made to lighting.

2:00 p.m. Venue concessions staff arrive and begin setting up.

3:00 p.m. Venue bar staff begin setting up. Merchandise worker sets up in the front lobby of the venue.

4:21 p.m. Instrument technicians re-enter venue space and begin checking instruments again.

4:38 p.m. FOH engineer returns with other members of the road crew.

4:44 p.m. Band begins soundcheck.

4:55 p.m. Band finishes soundcheck and returns to the backstage area. The TM, who arrived earlier with the artist, watches from the middle of venue floor; now talks with PM.

5:00 p.m. TM, PM and crew return to the backstage area.

5:00 p.m.–10:15 p.m. Crew members and artist are in the backstage area, which functions as a multipurpose facility. This time period represents an overlap between work activities and downtime for different members of the touring party. The TM and PM work continuously during this time, attending to emails, printing setlists, making hotel and food arrangements, holding meetings with venue personnel, and attending to artist's needs. The PM picks up coffee from a local coffee shop. Several crew members rest or take naps on chairs. Members of the touring party use the laundry machines. Several band members leave with a runner to visit the city. Two band members confirm the set list for that evening. The bus driver arrives from sleeping at the hotel and is in the dressing room area.

5:30 p.m. Catering arrives from a local restaurant. Crew and artist take turns eating.

FIGURE 3.1 (Continued)

7:45 p.m. –8:15 p.m. Support act performs.
8:45 p.m. Headliner begins. TM escorts the artist to the stage prior to performance, and watches part of the performance from a balcony.
10:10 p.m. House lights raised prematurely prior to artist's second encore.
10:15 p.m. Headliner finishes. TM escorts the artist to the backstage area/dressing rooms. Crew begins to dismantle the stage.
10:17 p.m. Local crew begin to take apart the audience barricade.
10:34 p.m. Backline technicians pack up instruments and risers are dismantled. Backdrop and lighting rig is lowered and starts to be dismantled. Crew members call out instructions to indicate the order in which cases should be loaded onto the truck.
10:39 p.m. Stage backdrop placed into a case. Local crew sweep the venue floor.
10:41 p.m. Last item removed from the stage and final cases remain to be loaded into the truck.

FIGURE 3.1 (Continued)

to and in preparation for a given show day, when TMs and PMs "advance" the show requirements with promoters and local crew members.

The content in Figure 3.1 is based on observations made at Venue B, which involved close observation of the entire process of a show day. This is not to say that the working day at Venue A was remarkably different. Load-in occurred, on average, between 9:00 and 10:00 a.m., with the earliest being 8:00 a.m. and the latest 2:00 p.m. Soundchecks also occurred in the late afternoon, usually around 3:30 p.m., with the earliest at 12:00 p.m. and the latest at 4:30 p.m. Concerts generally finished and the crew began packing up by 10:00 p.m. Taken together, working days at Venue A averaged 12 hours. Observations at Venue A also indicate that road crew members are often on-site before they actually start work activities. Being on-site can mean being in the bus parked outside of the venue, or involve going to catering for breakfast, taking a shower, or spending time in a dressing room prior to starting tasks for the day.[1]

A show day for TMs typically involves work activities that are related to the immediate concert as well as tasks for other events. As tours are necessarily planned in advance, "[l]ong before the first date of the tour much 'backstage' work is already done" (Weinstein [1991] 2000: 203). The planning that begins months ahead continues as the tour progresses and individual concert dates become closer in time. While TMs attend to the necessary tasks on a given show day, they also spend time preparing for the next ones. Much of the time working in their production office is future-focused and consists of advancing the details of upcoming shows by communicating with promoters, hospitality and hotels in other venues and cities – confirmations and adjustments to months of careful pre-planning and coordination. At the same time, they are overseeing the requirements of the day, processes that should go without incident if they have advanced accordingly and as such are entirely present, liaising with promoters, security and the relevant local crew at the current venue. The site of one concert effectively becomes a workplace for planning another.

Crew members have "clear roles and responsibilities" (Webber and Klimoski 2004: 267) and, as Figure 3.1 shows, specific road crew members are needed to perform tasks at different times. Start times follow a particular order as production elements build on each other. However, while crew members may begin at certain times, their activities overlap and multiple forms of work occur simultaneously. The daily schedule demonstrates that the length and intensity of the working day varies according to role. Load-in, set up and line check is a relatively routine and fast process, but is followed by several hours of relative inactivity prior to soundcheck and then the actual show. In this way, a show day is not characterised by continuous work but by duration over time; it involves periods of intense activity and downtime, and a key component is waiting.

When research respondents were asked about what they spend the majority of their time doing while on the road, several cited waiting as the most common activity. Craig (2017) described he was often "waiting for something" and Andy (2017) stated that "these days ... all I pretty much ever do is festivals, it's just waiting." Tony (2017) referred to waiting as "a big factor in my day" and specified that his job requirements involve "waiting for my boss to come out of his room, waiting for the show to start, waiting for guests to come, waiting for people to make decisions" (Tony 2017). Building on research I've published elsewhere (Kielich 2021), these comments from research respondents indicate that waiting is an essential aspect of the daily lives of many types of workers in the live music industry. The activity of waiting is amongst the characteristics of road crew members' jobs as being present, available and reachable are important components of their roles. Waiting also means that a crew member is ready to fulfil particular job responsibilities and can attend to unexpected issues that may arise. This aspect of a show day has been viewed with an element of mystery in the sense that it may appear as though workers are "walking back and forth apparently doing nothing" (Behr et al. 2016: 18). In actuality, walking back and forth indicates that they are in the process of fulfilling an important work activity (Kielich 2021: 121).

Scheduling is significant and has implications in the working lives of road crews. The time that crew members are needed to start working at the venue often relates to the amount of sleep they get. Generally speaking, those arriving early sleep less and those needed later get more rest. That this is an issue is expressed through humour, which is a strong aspect of the culture of touring and will be discussed in more detail later in Chapter 4. The backline crew is a particular target due to how they are the group of workers who are the last to arrive and who get the most sleep. For example, Adrian (2018), a guitar technician, usually starts at 10:00 a.m., finishes around midnight or 1:00 a.m. and goes to sleep around 2:00 or 2:30 a.m. As such, the backline is called the "country club" to imply a kind of leisurely, unhurried existence. Targeting specific members' working days and mocking them for being relatively easier than others illuminates the long-hours culture that is characteristic of tours. That crew members outside of the "country club" have bestowed such a nickname on the backline crew is also indicative of the ways they use humour to manage differences between them.

These features of the everyday workplace culture of touring can point to limitations in the academic literature, such as the usage of the term "backline" in Vilanova and Cassidy's (2019) article "Live Music's Backline Labor," which focuses on merchandise (or "merch") staff, specifically women, on tours. The article is limited by the fact that the authors do not define or explain what they mean in their use of "backline." It seems that theorising merch workers as "backline" is an attempt at an approach similar to that taken by Vicky Mayer (2011) in her use of the term "below the line" in her study of television workers. Mayer's (2011: 4) use of "below the line" follows industry conventions that separated workers "along a division of labour known as 'the line'." Below the line workers refer to those using "manual skills under the control of managers" and are understood in contradistinction to creatives and professionals (ibid.). This term applies to and can safely account for a wide range of workers due to the clear division in definition. Backline is an industry term that refers at once to equipment and the attendant technicians. The difference is that "backline" is highly specific and does not have the capacity to do the kind of theoretical work that Vilanova and Cassidy intend. To generalise this term is to do a disservice to the multiplicity of roles on a road crew and is a misuse of an operative industry term. Furthermore, the authors overlook the importance of the workplace culture in understanding and theorising what it means to work in live music. Their use of the term "backline" is incompatible with the connotations of the colloquial "country club."

During a show day, numerous other activities accompany work or occur during lulls in activity. Road crew and local crew members stand around and talk while waiting for their next activities, the FOH engineer talks with others and attends to guest list requests from members of the local crew. During the process of checking the sound and doing the line check, crew members engage in banter and humorous exchanges. Observations also highlight anomalies that can occur during a show day and how schedules are adjusted accordingly. The minimal activity in the afternoon at Venue B was due to the fact that the band and crew wanted to watch a major sporting event on television, which they did in the dressing room. The schedule of the day was adjusted to allow time to accommodate watching the game and also created a longer lull of activity in the day than was typical. It is additionally useful for understanding the waiting that can be experienced throughout the day as well as the variations that can occur in order to accommodate interests and create normalcy.

Despite the lengthy hours, crew members reported feeling as though they do not have enough time to finish their tasks. Adrian (2018) referred to the need to work to a deadline as "possibly the most stressful part" of the job:

> It's just knowing that there's that deadline which is showtime and everything has to be ready. That can be stressful depending on what you've been doing earlier in the day ... whether there's a knock-on effect of any delays, cause we're last in the building we usually feel the worst pressure for time delays cause shit rolls downhill.

Adrian's quote suggests that the backline crew experience pressure based on their later arrival in the organisation of a show day. This runs counter to the perceptibly relaxed working day they are teased for via their nickname. While they may be advantaged by a later start and more rest, they are disadvantaged by a shorter period of time before the concert and the potential for trickle-down problems to complicate their working practices.

> ... I guess cause of the time restraints and what has to be done the whole day is stressful but you just get used to that stress and it's just part of the day. It's possibly wrong to characterise it as stress, but there's always an awareness of the time restraints that you have to work with so sometimes you have to prioritise tasks, sometimes you have to postpone tasks and get through it. I think if you speak to enough guitar technicians every single one of us has gone through a show knowing that something is broken.

Adrian's comments indicate that show days are marked by unpredictable circumstances and situations and suggest that time constraints are experienced differently by the various departments of road crews at the same time that they underpin all of their working lives. Time constraint can also affect the ability of crew members to properly attend to important aspects of their jobs, which creates risk in the management of their reputations. The reporting of consistent time pressures further implies that stress is taken as a normal and consistent feature of a show day and as something to be managed.

Research respondents cited the number of hours as a major component in the day-to-day experience of working on a road crew. Amy (2018) described her working day by "you just don't get to put it away at 5 p.m." Head of security Tony (2017) summarised the hours as follows:

> The hours are definitely probably the biggest factor. A normal typical working day worldwide I would say would be somewhere between 8 to 10 hours. That's a half day for us for the most part. I will regularly be working 18 hours a day, seven days a week for months at a time.

Road crews are like other types of behind-the-scenes workers in that both "invest untold hours" (Curtin and Sanson 2017: 1). A long-hours culture is a consistent attribute of working in the cultural industries. As example, David Hesmondhalgh and Sarah Baker (2011: 116) observed working patterns of ten hours or more per day, seven days a week, among some workers in the television industry. Respondents gave further indication of the long hours, as well as their significance, by positioning this characteristic as a determining factor for new entrants to the field or potential crew members. Craig (2017) and Adrian (2018) both stated that touring is not an environment for anyone who is not willing to work "ridiculously" or "very very" long hours and noted that an absence of such motivation acts as a filter. These findings suggest that road crew members accept these conditions as part of the expectations of their jobs and among the non-negotiable features of the everyday life of live music. Research findings also suggest that TMs in particular are attracted to and motivated by this line of work because of how their own personalities match the long hours, the demands of touring and the specific requirements of their positions. Joe (2018, 2019) described himself as constantly "up and moving ... I really like being busy. I'm very bad at being idle." TMs characterise themselves as having a certain restlessness, a preference for being occupied with activities and an intense work ethic. In her book, Claire Murphy (2019: 20) justified her suitability for tour management, which is not her primarily role on tour, by how she is a "pretty organised person and loved to work 24/7." Amy (2018) stated:

> ... I think that everybody who does this has a pretty extreme work ethic, like you will work until work gets done. And it's like people are not always, myself included, not always so good at knowing when to stop and that's why you see people burn out, you know and people are now talking about 'you gotta take care of yourself' and ... you just go until the work's done, like people can be pretty intense. I think that it's like people are pretty perfectionistic, but ... not to the point where it necessarily stops them from doing their job, but to get everything, you know, it's a point of pride to get everything exactly right and have it work like clockwork.

The long-hours culture is further evident by the ways in which work activities transcend the boundaries of designated work days. Road crew members label touring schedules in terms of "show days," "travel days" and "days off." For TMs, the latter can be characterised more precisely as "non-show days" due to how they often still need to work on days on which there is no live music event (Amy 2018). The responsibilities involved in tour management (see Chapter 5) have been characterised as "oppressive and alienating," "consuming" and that the "pressures applied from all directions" come with the potential to reduce TMs to tears, albeit that must be reserved for "behind closed doors" (Hawes 2019: 344–345; see Chapter 4). Kim Hawes (2019: 345) described longing for periods of sleep, or that "period of sweet unconsciousness when I didn't have to think about everything I had to do" and makes a point to warn potential entrants to the field of this aspect of the job. The TM at Venue B expressed that he was looking forward to having the following day off in order to attend to emails and get caught up on work. A non-show day provides time and space to work away from the demands and time constraint of a show day.

For road crews, the long-hours culture in which they work is the product of the overall length of a tour, the repetition of this pace over an extended period of time and the steady mobility of travelling between dates. An important aspect to understanding the working hours of crew members is that the nature of touring means that these conditions are continuous and sustained throughout the course of a tour. In addition to the hours involved in a single concert, work also transcends spatial boundaries as it occurs in venues, on forms of transportation and in hotel rooms. The following section looks at one such space, the production office.

The Production Office

The production office is any space within a venue in which the TM and PM, and assistants if applicable, work for the day. The rider indicates the number of tables, chairs, outlets and internet connection needed. The production office is the site of the planning and administration of the tour, with the efforts of workers being applied to both the immediate tasks of the day and to advancing upcoming shows. Most venues have available space that is either specifically reserved for or can function as a production office. If a venue cannot provide adequate office space, makeshift spaces are often utilised as production offices. As TM Andy Reynolds (2008: 31) observed, production offices sometimes take "the form of a space in the cloakroom with a desk and a telephone[.]"

The mobile nature of touring means that TMs are prepared for these conditions. Tour and production managers generally have portable production cases that contain the necessary equipment, including computers, printers, office supplies, credentials and paperwork, needed for advancing and running the tour. They are essentially smaller versions of a musician's flight case that contain their wardrobe and personal effects. TMs and PMs refer to these cases as their "office." At Venue B, the PM explained the delay in providing a credential because he did "not have his office yet," meaning the production case had not been unloaded from the truck and placed backstage. Live music managers treat their "offices" as people do in conventional ones by placing photos, decorations and other personal touches within them, and how they do so – items from popular coffee chains, badges that reference the roles of crew members, and the use of alcohol-branded merchandise as organisational tools — shows their embeddedness in the culture of music and the lifestyle of touring. However, though these cases are portable, they often require desk space to unpack some of the items in order to facilitate work for the day. Regardless of the nature or type of space, the presence of the production team and their supplies means that these spaces become temporarily reimagined as the production office. Makeshift spaces that crew work in and, to some degree, are accustomed to are both evidence of and reproduce the highly informal nature of working in the live music industry.

When space is a commodity, musicians' comfort and privacy are prioritised with access to dressing rooms while the workspaces for TMs and PMs can be compromised. At Venue B, the TM and PM had to set up their offices at tables in the closely shared and busy space of the green room, which also contained catering and was the major hub of activity during the day. In this way, it can be said of the touring workplace that artists are less likely to have to adapt to their environments due to the efforts of their staff, but those who create those possibilities must be highly adaptable themselves. Even prior to arrival at a venue, the cost of touring, and the responsibility of the TM/PM to manage the budget, affects their workspaces. In order to save money on freight costs, TM/PM Joe (2018) willingly chose to take the "absolute minimum" work supplies needed in order to fit them into a smaller pelican case rather than the usual production case. He also limited what the guitar technicians and FoH engineer could bring, all of whom were unhappy but understood. Though artist management was in support of paying extra fees, he said "this is an awful lot of money and I knew we could make it work." In this way, being a "support" worker also means forgoing individual needs and taking measures that benefit the larger interest of the tour. Being a support worker involves privileging the working needs and comforts of superiors. It involves an expectation to be adaptable so that those being supported are less likely to have to compromise.

At Venue A, the informality that is a normal part of working on tour was in some ways exacerbated due to the multipurpose function of the facility. A room that was used to store tools was frequently assigned as the production room due to the adequate desk space. The unusual overlap of occupational space was not lost on the visiting crews, who would make jokes about using the tools. The local crew, which is the group of workers at a concert venue who assist the touring party with all aspects of the live music event (see Kielich 2021: 120), tried to remove or cover some of the equipment, and offered apologies, but little could be done to alter the combined workspace.

The humour that crew members use to temper these situations highlights how accustomed they are to adapting, the importance of their willingness to compromise in order to meet work goals and is also part of a wider cultural norm of coping with the conditions of touring. The use of humour to overcome, or mask, any frustrations about deference or compromise in the workplace illuminates the "interrelationship between humour and masculinity in the social relations" of touring (Collinson 1988: 181). Doing so takes on additional significance in a male-dominated workplace given that these behaviours and expectations are stereotypically deemed feminine.

Despite the production office's mobile nature and potential for compromise, it is a central media operation and communication hub, or "nerve center for the tour day" (Reynolds 2008: 31). Matt McGinn (2010: 83), a long-time guitar technician for Coldplay, characterised the production office as "an inner chamber of extreme professional masochism" based on the willingness of the production team to handle their many responsibilities and attend to a range of queries. Within a concert venue, it is the source of important documents and information that help to orient workers within the venue and assist with communication between the road crew, local crew and other relevant parties.

Amongst the most essential documents is the *rider*, which is attached to the contract agreement between an artist and a promoter (Workman 2012: 119). I have published research on the rider elsewhere (see Kielich 2021) and, as such, will be brief here. As I have theorised, the rider is an essential component of a tour's supply chain (Kielich 2021: 116) and a guiding component of a show day. The rider details logistics and provides specific information regarding the needs of the artist and crew. It includes the technical requirements for the concert, such as the instructions for how the stage, lighting and sound should be configured, along with any equipment that must be hired locally. It also includes sections that indicate the food and drink requirements for musicians and crew. TMs and PMs must ensure that the specificities outlined in the rider are met and adhered to with the local crew on a show day. Most riders are made available to the promoter and local crew months in advance of a concert, but updated versions are often sent closer to an event and further changes circulated three days prior to a show day (122). In addition, they "advance" the rider requirements for upcoming shows during the working day. TMs state that, aside from unexpected issues, a show day should go smoothly if everything is properly advanced. Part of this process depends on the nature of the communication between the TM and the local crew member in advance.

The *day sheet* is a document that outlines the schedule for a show day and orients workers to their location. A reading of the day sheet's contents illuminates the information and activities deemed most essential to making a concert happen. It is a structuring document that communicates when and where workers are needed. In general, a day sheet provides details such as the date, city and venue and the scheduled times of all relevant events for the day. This includes transportation and physical arrival at the venue, the set times for load-in, soundchecks, door and performance times and curfew and bus call times. Furthermore, mealtimes and types (breakfast, lunch and dinner; catering or buy out) are included. Additional information, such as the next city and venue, the distance and time to travel there and hotel information, may also be provided, but the content is based on the preferences of the TM (Workman 2012: 231). A particularly detailed day sheet generated by a TM at Venue A also included the type of venue, the weather forecast, contact names for the promoter and local production manager and internet access information. Furthermore, at the bottom of the page, it featured a "word of the day" which shows the efforts TMs make to enhance morale. An essential component of the TM's job is to maintain a "constant flow of information to [the] organization" that is a major factor in the tour's success (ibid.). Doing so means that all personnel have accurate information and understand what to do and where to be (ibid.). In this way, the day sheet is also a reflection of the quality of an individual TM (or PM). TMs or PMs print the day sheets in their production offices and post them in the dressing rooms, green room and bus. The day sheet is also posted in the local production office to inform local crew.

Additional documents may also need to be generated in the production office based on the specifics of a venue. Most TMs have laminated, reusable *signs* that they use to direct musicians and crew to their relevant places.[2] For example, the principal artist and band typically have signs for each of their dressing rooms, and one would also be made to label the production office. Signs are also used to guide workers to the areas in a venue where these rooms can be found. They are generally placed in the venue as soon as personnel arrive and before the artist is onsite. The use of signs is intended to ease the TM's workload and prevent frustrations on the part of the crew and musicians. Workman (2012: 171) equates signage with professionalism and notes that not using them results in wasted time explaining directions and locations to musicians and crew "that they can't afford to waste" and, in some cases, do not even have enough information to do.

The importance of the communicative nature of these signs is clearly demonstrated when it fails. During participant observation, an assistant TM, working for a tour with multiple artists on the same bill, incorrectly labelled two of the rooms, which were at significant distance from each other in the venue. This error initially created confusion, and the potential for extra work, for the hospitality team when rider items were being distributed and arranged. On another show day, the layout of the venue presented problems for a band member who was allocated a private dressing room in a different area. Despite the signs already being in place, he could not locate them and was initially unable to figure out where he was supposed to be. In this way, ready-made signs do not always suffice. In another instance, Suede bassist Mat Osman (2023) posted a photo on his Twitter account of a confusing arrangement of signs in the backstage area. The signs were intended to indicate the respective directions towards the locations of Suede's catering and the stage. However, the directions of the signs' arrows, in some cases, were contradictory and, in others, were covered up with other signs. Osman included the caption "Still have no idea where catering is. Send supplies" which humorously confirmed his inability to successfully navigate the backstage area.[3] The configuration of Venue A led to TMs/PMs printing special signage to assist with navigating the backstage area, including arrows or additional "to catering" signs. Despite the usage of signs in the backstage area at Venue A, they were not permitted in the main part of the building. As such, given their local knowledge, the responsibility of explaining whereabouts, and physically walking with members of the touring party to catering, was handled by the local crew.

Aftershow food orders are another aspect of live music that generate a variety of handwritten and printed documents. Usual practice during participant observation was that the hospitality staff would leave a folder of menus in the production office and communicate with the TM or PM later in the day to receive the order. TMs have individual methods of communication for aftershow food. Some prefer to text the details or call with the information. Others provide a handwritten document organised according to band, crew and/or individual names of personnel. Several, usually those working for more experienced groups, had their own Word templates that featured the name of each person in the touring party followed by the respective food order. These templates also gave instructions regarding the labelling of food and requests for condiments, utensils and where and when the food should be delivered. For an example outside of participant observation, at the start of the North American leg of R.E.M.'s 1995 *Monster* world tour, TM David Russell circulated a document to each of the four band members requesting aftershow food preferences, which he could then refer to throughout the tour. The template included their name, beverage requests, pizza preferences and the frequency with which they were comfortable eating it, sandwich requests and any other acceptable items. Russell added introductory comments that carefully stated to "remember that in some places we will be limited as to what form of heartburn is available at 11:00 p.m." and reflected the humour specific to the culture of touring that was ubiquitous in his communications (REM Aftershow Food Preferences Template Monster World Tour 1995, David Russell Collection). Russell's approach to handling aftershow food is indicative of the challenges of coordinating personal preferences with the realities of a local area. More than this, however, it shows that efficient and advance planning, and efforts made to manage the expectations of musicians, can ease the daily lives of tour managers.

Aftershow food generally needs to be placed into the dressing room, production office or bus, prior to or just after the concert ending, which creates a particular time frame for it to be ordered and picked up. The volume of food requires orders to be placed with sufficient time for preparation and pick up, and it should still be warm (if applicable) upon arrival at the venue. Hospita lity staff generally placed the food order via phone from within the local production office. Simple orders, such as pizza, did not create problems and were a regular choice for TMs due to ease and limited room for error. More complex orders came with challenges, such as requested items no longer being on the menu, the restaurant running out of certain items or not enough information being provided by the touring party regarding preferences for side items or toppings. In such instances, the hospitality staff had to locate the TM on the premises and subsequently reiterate the revised information to the restaurant. This could result in delays in food orders which had a domino effect on the rest of the schedule. Another challenge of aftershow food was with its distribution. Requests were sometimes made to restaurants for the individual names of musicians and crew members to be written on food boxes, which was not always respected. Even when it was, the process of locating specific boxes and organising them for delivery to the appropriate party was time consuming and usually needed the assistance of two or more people.

The ordering of food also revealed that artists can take priority over crew, and that support personnel compromise in musicians' favour or attempt to manage their expectations. One TM was unsure as to the exact request for a band member and ordered a couple of items in his name as backup, stating that he could just have the other one depending on the band member's choice. In another case, the TM completely forgot to order aftershow food for the crew, resulting in their food being delivered much later. The artist is sometimes given more choice and specificity while the crew is generalised. For example, a sushi order included variety and an itemised list for the band, with a special request for extra condiments. For the crew, the order was "one sushi boat" to be shared. Several tours had environmental concerns that affected preferences around food and drink packaging. A TM who made special requests for food packaging when ordering from a restaurant was concerned with ensuring that the singer's food arrived in the type of packaging he expected. For the crew, however, any mishandlings "could be dealt with." This suggests that as long as the singer's packaging arrived intact, it gave the impression that everyone else's did, also, which likely had less to do with the environment and more to do with keeping the peace and maintaining a reputation.

These daily communications demonstrate that the various documents generated by a concert event "constitutes the event … of which it is itself a part" and structure the identities of personnel (Prior 2003: 72, 108). As such, these documents function as key social actors (Latour and Woolgar 1986, Latour 1990, Riles 2006) that help produce live music events and are integral components to the working lives of road crew members in the context of a show day. A summary and conclusion of this chapter follows.

Conclusion

This chapter has discussed the various features of a show day in relation to members of road crews. It has outlined their responsibilities and the ways in which they occur and overlap, including the integral activity of waiting. It has also demonstrated that the working lives of crew members are characterised and shaped by a long-hours culture. The latter is the product of the length of a show day, which is formed by the many activities that must be enacted. Following this, concerts are realised by a long-hours culture as much as they are also the cause of these working conditions for crew members.

Exploring the temporary workspaces of TMs and PMs in the form of the production office indicates that being a support worker involves privileging the working needs and comforts of superiors. It emphasises an expectation to be adaptable so that those being supported are less likely to have to compromise. TMs encounter vastly different spatial configurations, and their responsibilities are enabled or limited by the resources of a local area, which requires troubleshooting and consideration. In this way, the chapter shows that efforts involved in attending to the needs of artists reproduce the special privileges they are accorded (Becker 1982). The number of labour hours involved in realising a live music event further substantiates this factor.

The chapter has also addressed the diverse set of documents that crew members utilise and rely upon throughout a show day. It has provided insights into the usefulness of best practices in relation to the communicative function of these documents. Their functionality and importance are most clearly understood when they fail, which emphasises the need for organisation in a mobile workplace. The next chapter will examine the culture and experience of being on tour for members of road crews.

Notes

1. See Webster (2015) and Gorman (1978: 20) for additional insights about the working day of a show day.
2. Outside of the backstage setting, Webster (2011: 215–221) has shown how visual and aural signage plays a significant role in managing audience behaviour at concerts.
3. See https://twitter.com/matosman/status/1639319716303863841

References

Becker, Howard S. 1982. *Art Worlds*. Berkeley: University of California Press.

Behr, Adam, Matt Brennan, Martin Cloonan, Simon Frith and Emma Webster. 2016. "Live Concert Performance: An Ecological Approach." *Rock Music Studies* 3(1): 5–23.

Collinson, David L. 1988. "'Engineering Humour': Masculinity, Joking and Conflict in Shop-Floor Relations." *Organization Studies* 9(2): 181–199.

Curtin, Michael and Kevin Sanson, eds. 2017. *Voices of Labor: Creativity, Craft, and Conflict in Global Hollywood*. Oakland: University of California Press.

Gorman, Clem. 1978. *Backstage Rock: Behind the Scene with the Bands*. London: Pan Books.

Hawes, Kim. 2019. *Confessions of a Female Tour Manager*. Independently Published.

Hesmondhalgh, David and Sarah Baker. 2011. *Creative Labour: Media Work in Three Cultural Industries*. New York: Routledge.

Kielich, Gabrielle. 2021. "Fulfilling the Hospitality Rider: Working Practices and Issues in a Tour's Supply Chain." In *Researching Live Music: Gigs, Tours, Concerts and Festivals*, eds. Chris Anderton and Sergio Pisfil, 115–126. London: Taylor & Francis/Routledge.

Latour, Bruno. 1990. "Drawing Things Together." In *Representation in Scientific Practice*, eds. Michael E. Lynch and Steve Woolgar, 19–68. Cambridge, MA: The MIT Press.

Latour, Bruno and Steve Woolgar. 1986. *Laboratory life: The Construction of Scientific Facts*. Princeton, NJ: Princeton University Press.

Mayer, Vicki. 2011. *Below the Line: Producers and Production Studies in the New Television Economy*. Durham, NC: Duke University Press.

McGinn, Matt. 2010. *Roadie: My Life on the Road with Coldplay*. London: Portico Books.

Murphy, Claire. 2019. *Girl on the Road: How to Break into Touring from a Female Perspective*. Independently Published.

Osman, Mat (@matosman). 2023. "Still Have No Idea Where Catering Is. Send Supplies." Twitter, 24 March, 5:34 p.m. https://twitter.com/matosman/status/1639319716303863841?s=46&t=u96kSXoF9N4ClGW6M-KUPQ

Prior, Lindsay. 2003. *Using Documents in Social Research*. London: Sage Publications Ltd.

REM Aftershow Food Preferences Template Monster World Tour 1995, David Russell Collection, Library and Archives, Rock and Roll Hall of Fame and Museum.

Reynolds, Andy. 2008. *The Tour Book: How to Get Your Music on the Road*. Boston, MA: Cengage Learning.

Riles, Annelise. 2006. *Documents: Artifacts of Modern Knowledge*. Ann Arbor: University of Michigan Press.

Vilanova, John and Kyle Cassidy. 2019. "'I'm Not the Drummer's Girlfriend': Merch Girls, Tour's Misogynist Mythos, and the Gendered Dynamics of Live Music's Backline Labor." *Journal of Popular Music Studies* 31(2): 85–106.

Webber, Sheila Simsarian and Richard J. Klimoski. 2004. "Crews: A Distinct Type of Work Team." *Journal of Business and Psychology* 1(3): 261–79.

Webster, Emma. 2011. "Promoting Live Music in the UK: A Behind-the-Scenes Ethnography." PhD diss., University of Glasgow.

Webster, Emma. 2015. "'Roll Up and Shine': A Case Study of Stereophonics at Glasgow's SECC Arena." In *The Arena Concert: Music, Media and Mass Entertainment*,

eds. Benjamin Halligan, Kristy Fairclough, Robert Edgar and Nicola Spelman, 99–109. New York: Bloomsbury Academic.
Weinstein, Deena. (1991) 2000. *Heavy Metal: The Music and Its Culture*. Cambridge, MA: DaCapo Press.
Workman, Mark. 2012. *One for the Road: How to Be a Tour Manager*. Road Crew Books.

Interviews

Adrian. Guitar Technician. In-Person Interview, 24 May 2018.
Amy. Tour Manager. Skype Interview, 9 July 2018.
Andy. Tour Manager and Sound Engineer. In-Person Interview, 18 April 2017.
Craig. Monitor Engineer. In-Person Interview, 4 May 2017.
Joe. Tour and Production Manager. In-Person Interview, 14 May 2018. Skype Interview, 28 August 2019.
Tony. Close-Protection Security, Head of Security. Skype Interview, 4 August 2017.

4

BEING ON TOUR

Touring is the central activity in the working lives of road crew members. As stated in the Introduction, touring – or being on tour – refers to the experience and practices involved in realising a sequence of live music events. Live music is a structure that distributes important roles and duties, and people occupying these roles are part of a complex mobile world. On tour, members of road crews encounter particular conventions and conditions that structure and shape their working lives. They also participate in a workplace culture that enhances the ability to co-exist cohesively and effectively facilitate the realisation of live music and that, in turn, fosters a particular way of life (Williams 1983: 91).

Road crew members define, perceive and experience touring as distinct from the "real world" that exists off the road. Research respondents refer to touring with such terms as "a bubble," "outer space" and "our world" and differentiate themselves from the "civilians" who do not work on tour (see also Gorman 1978: 15). Tana Douglas (2021: 119) described the experience of touring as "having no idea what's going on in the word outside that insular orb, and you tend not to care." Louise Meintjes (2003) and Antoine Hennion (1989) took a similar perspective, and used corresponding terminology, in their studies of the recording studio – studies that are useful for conceptualising the road crew's sense of the world of touring. The recording studio is "an enclosed space … a private space with a secret life" (Meintjes 2003: 84). It is "cut off" from the "outside" or the "real world" and is "made to the measure of people so that they can test their own creations" (Hennion 1989: 407). This perspective is further reflected in tourism studies. Buses, a common mode of travel on tour, transport "passengers in a closed environment" that creates a "physical and psychological distancing … from their outside environment"

DOI: 10.4324/9781003303046-5

(Holloway 1981: 381). By temporarily leaving "the real world," the world of touring becomes a compartmentalised space. This separation is further enhanced through the manner in which crew members adapt to the road, the specificities of which substantiate that to be "on tour" is as much a mindset as it is a mobile activity and workplace.

This chapter is about the experience of being on tour and the characteristics of the complex mobile world of touring. It shows how touring shapes the lives of road crews and is a significant component in the realisation of live music events. This chapter demonstrates that the conventions and culture of touring define working life for road crew members. It begins by addressing aspects of the social experience of touring and how road crew members manage working and living closely and continuously with others. From there, it analyses the disorientation that can and does occur on the road, which is linked to a tour's itinerary and is a common stereotype about the experience of touring. It continues by exploring the camaraderie and friendship that develops between crew members and moves to analyse the ways in which masculinity underpins and operates within the workplace culture. The chapter then draws attention to the ways that touring can affect crew members' mental and physical health. The chapter then discusses how members of road crews adjust to life at home when they are off the road.

Working and Living Together

Being on tour positions road crew members in an intense setting in which they work and live in close proximity for extended periods of time. Crew members state that they "live on top of each other" or "in each other's pockets" to describe this experience. The close nature of the touring environment is one of the reasons why the hiring process, discussed in Chapter 2, relies so heavily on contacts and referrals.

> If you're gonna be with a bunch of people 24/7 ... it's a lot of vibe, feeling like you're a good fit, and it's like how else do you know unless you've been with that person, unless you've actually worked with them?
>
> *(Amy 2018)*

The need for road crew members to travel between dates on an itinerary, and the forms of transportation they take when doing so, creates the close proximity on tour. Mobility is how "geographic movement becomes entangled in the way societies and cultures assign meaning through talk ... and live out their lives" (Cresswell 2006, see Adey 2017: 7).

The mode of transportation in which crew members travel varies with the scale of the tour and resources of the artist. Two research respondents who work for non-mainstream rock and folk artists travelled most frequently by van; the majority of respondents work for established rock musicians and travel in a tour bus, also called a sleeper bus; and one respondent works for a top-level rock band and travels by private jet. During participant observation at Venue A and Venue B, the majority of headliners and their road crews travelled by bus, though in some cases, cars were used to transport members of touring parties from hotels to the venue, leaving it unclear what the primary mode of travel was between cities. Those who travelled by van tended to be support acts.

The specific mode of transportation that road crews utilise creates conditions and shapes their experience. Different forms of transit may facilitate more or less intense interpersonal situations and may enable, constrain or define social interaction. Van touring is usually associated with a smaller division of labour and more acute social experience. A typical van used on a concert tour fits up to eight people, has limited spatial comfort and a complete lack of privacy. They do not feature conventional sleeping accommodations nor convenient access to toilets. Tour buses, or sleeper buses, are the "standard travel method for modern tours" (Reynolds 2008: 350). Sleeper buses are typically occupied by 10–14 people (ibid.). They are essentially standard buses that are converted and are equipped with sleeping accommodations in the form of bunk beds, which are typically stacked in threes on either side of a corridor. Sleeper buses also feature a bathroom, shared eating and living spaces and lounges at the front and back.

Travelling by plane represents less intense proximity and more opportunities for comfort and privacy. Plane travel tends to be reserved for the band, its entourage and key management and touring personnel, while other members of the same touring party travel by bus. Who travels by plane is based on a combined factor of role on the tour and seniority in the artist's touring party. With a different artist, some personnel on the plane would usually be on the bus with the rest of the road crew. As such, the plane represents a hierarchy that is inconsistent with the norms in the road crew division of labour, and, as a result, the topic of who travels by plane is "quite divisive" (Tony 2018). Crew members who "hub," as explained in greater detail later in this chapter, go to the airport after a show, fly to the hub destination, which is usually four hours or fewer away, and stay the night at a hotel. An in-depth comparison of the experiences of travelling on different modes of transportation is beyond the scope of this book. For the purposes of this study, the contrast between them is less significant than what they have in common: they position crew members in settings that extend their proximity to others outside of the scope of a show day, and, though transient, they provide a "framework for social interaction" (Holloway 1981: 398).

Furthermore, being on tour is ultimately to be mobile and to simultaneously construct a sense of home, which these various spaces come to function and represent. Road crews are at once away from home at the same time that they exist within a temporary one formed by and within several types of spaces, including tour buses, hotels and venues. Rather than one specific locale, a "home" is an "active practising of place" in which familiarity is "actively produced over time" (Felski 1999: 24, see also de Certeau 1984). Home is also defined by, and a site in which, power struggles and inequalities related to gender, class and generational differences (ibid.) are negotiated.

The confines of living on a bus create particular expectations around conduct, a fact which contributes to both a more agreeable and safer living environment. Reflecting on the experience of her first tour, Claire Murphy (2019: 14) highlighted that she "had absolutely no idea what I was doing" and that she "didn't know the etiquette of a tour bus." Understanding how to live on a bus is an essential component to working on tour and is something that is learned.

Several essential aspects of living on a tour bus were highlighted during research. The close-knit bunk bed sleeping arrangements depend on keeping noise levels down while others are asleep. When the bus is parked, it is expected that it is kept locked to safeguard personal belongings. Furthermore, road crew members are discouraged from bringing unfamiliar people onto the bus. Research respondents and primary source materials repeatedly expressed that the "cardinal rule" of bus touring is that "no solids" are allowed in the bus toilet (Joe 2017). This is due, in part, to the close proximity of bus life but has a practical reason in that bus toilets require maintenance too often, and this is too expensive (Cody the Roadie 2020). If someone violates this rule, it is usually the bus driver who must clean the toilet, and the imposition is unfavourably regarded. When crew members are unable to wait until the bus stops and they can use proper facilities, they must engage in a practice called "hot bagging."

Close working and living conditions place a particular emphasis on social cohesion and managing relations between road crew members. The nature of their working lives means they move between groups and must quickly adapt to new and different people or resume interaction with others. In the short term, doing so creates a more tolerable workplace while on tour. In the long term, it influences how crew members evaluate each other and maintain their reputations. The significance of social relations is expressed when road crew members refer to their touring colleagues as "family" for the kinds of bonding and shared, relatable experiences that life on the road creates between them. The term can also be understood to represent the tensions that characterise the nature of those relations, which is largely a product of the closeness and familiarity they encounter.

At the most basic level, relations between crew members are maintained by mutual respect and tolerance for others and their habits. Crew members must respect their colleagues and their personal working and living areas, attend to the cleanliness and maintenance of their own personal spaces and habits, refrain from creating difficulties and be cooperative and considerate. Given the variety of people's personal habits and preferences, doing so is not always easy. Crew members must "make yourself a vital part of the bigger picture," and adaptation and efforts made benefit the greater good of the tour (Adrian 2018).

> So you might find someone's personal habits slightly offensive, but if you … just leave it alone and shut off … if someone doesn't change their socks as often as they should, you know, maybe he has personal hygiene that you find not quite up to your standards, unless it's utterly offensive, then you just leave it. You have to, you have to get on.
>
> *(Craig 2017)*

Having the right attitude and a willingness to adapt are among the ways that road crew members "fit in" with the larger group on tour, which is integral to its functioning. Acquiring work is strongly dependent on the ability to fit in, and those who cannot fit in often do not continue working on road crews. "It's such a tight, closed work environment that you can't have people who don't play nice with others involved" (Tasha 2018). Fitting in can rank higher than skillset in terms of suitability for working on tour, which suggests it is not only integral to the immediacy of a given tour but also functions as a criterion by which crew members assess each other for future positions. Mary Ann Clawson (1999: 207) observed that "band members may become indispensable, and thus powerful, because of their business acumen, social skills, or long-time familiarity with the band repertoire; thus, the status of musician within a band is not fully predictable simply by knowing she is a woman and/or bassist." In the same way, road crew members may be valued for much more than their skillset and possess attributes that are desirable for how they function and what they contribute outside of the exclusive structure of a show day.

> … if people don't have good attitudes they get weirded out and they don't end up lasting on the road very long. I mean you can only, like your technical ability might be amazing but if you're just not fitting in with the crew, if you just don't fit in or working on the road is just not good for you, people tend to fall out, like they tend to not last long. And like attitude is kind of a huge, I would say attitude is way more important than technical ability quite often, just being able to fit in and play nice with other people.
>
> *(Tasha 2018)*

A major component in how they fit in is their demeanour. Crew members have been noted for their "curious enthusiasm, energy, dedication and general can-do attitude" (Ames 2019: i). In this way, they represent a positive attitude despite potentially uncomfortable or challenging working conditions. Such a demeanour is part of the norms that function as "the informal rules that groups adopt to regulate and regularise group members' behavior" (Feldman 1984: 47). Norms indicate expectations for behavioural conduct and express the group's values and identity (Durkheim 1983). Introversion or "standoff-ish" personalities could produce difficulties and are viewed as dispositions inconsistent with the norms of touring given the close contact and need to get along well with others (Adrian 2018). This need reflects the closeness of touring at the same time that it is an effort to manage it. To be "on tour" is not only to occupy a role and a set of responsibilities, it is a way of being and presenting oneself.

> ... on tour you never completely relax ... But because you're pretty much with other people on the tour 24/7, if you're sleeping on the bus, you can never completely switch off you know from whatever you're on-tour persona is, you've got to keep that on all the time.
>
> *(Craig 2017)*

The reference to an "on-tour persona" suggests that crew members adjust and modify aspects of their personalities in order to fit in with the group. This persona differs from the person they are and the attributes they may reveal off-tour. Tasha (2018) attested to this by observing that "what's really strange is when you hang out with people on the road very often they're different off the road." This suggests the act of impression management, or performance, put forth by Erving Goffman (1959). Impression management is the attempt by individuals to control the impressions that others form about them (Leary and Kowalski 1990: 34). Crew members put on a "front" that is shaped by their roles, social norms and context (40, 41; Goffman 1959). Impressions are strongly tied to the ways that others treat and evaluate each other and are an important motivational factor in the achievement of goals (38). Given the role of networking and referrals in the acquisition of work for road crew members, their reliance on others places particular significance on impressions (ibid.).

The on-tour persona also highlights that being on tour means crew members are always "on" given the norms and expectations associated with life on the road, as well as the importance of following them to acquire future work. This factor symbolises the all-encompassing experience of being on tour and suggests that the "working day" for road crew members is continuous. Goffman's (1959: 106) notion of "regions" further addresses this feature of their working lives. In a "front region" – the touring workplace – workers attempt to convince the audience – their co-workers and employers – that an "activity in the region maintains and embodies certain standards" (107). A "backstage region" is where a performer is supposed to be able to relax and refrain from the character. It is also an environment where the audience is not permitted access, which is a requirement in order for the performer to be able to prepare for the performance (111–114). Road crews have limited access to a "backstage region" in the context of their workplace. Though they spend a portion of their working day in the backstage area of a concert venue, it is among their audience: co-workers and musicians. The tour bus equally provides minimal privacy, with communal areas for relaxing and curtains separating bunk beds. As such, road crews have limited ability to acquire or inhabit backstage areas (114–115) where they can disengage from expectations.

These kinds of strains along with continuous and close living and working conditions mean that the touring environment functions as a type of pressure cooker. Kim Hawes (2019: 27) recalled first learning about the "dynamics of life" on tour. In particular, she observed that the camaraderie that characterised the start of a tour gradually transitioned into "bickering and bitchiness" and "outright belligerence" that resulted from continuous close quarters. While mutual respect and social norms are strategies designed to maintain order, personal limitations inevitably expire. Everyday occurrences that would otherwise not become issues have the potential to become exponentially magnified. Or, as Tony (2017) succinctly summarised, "dramas become real dramas." Musicians describe the social realities of touring in similar terms. Lol Tolhurst (2016: 169) of The Cure stated that "being in a touring band is rather like being married to the people you work with … You live with each other 24/7 for months on end, and the smallest irritant can become a very large dispute if you're not careful" (Frith et al. 2019: 172).

Friction can develop based on repetition and familiarity in which someone may "cause the same problem, the same joke everyday at the same time" (Joe 2017). A crew member may lose patience with another person's habits, demeanour or the overall pressures of touring. Hawes (2019: 27) links tensions between crew members with the conditions of travelling. In her case, she cited pervasive cigarette smoke and clutter left on the tour bus as contributing factors.

> I think that people have blowouts … it just becomes too much … you might [find] something about someone you find really annoying but you don't just … leave it. But then you know sometimes if someone keeps pushing your buttons and pushing your buttons eventually when they, on the bus when you've had a wee too much to drink that'll explode. And you know I have seen that on more than one occasion turn into a physical fight. And you know, that generally will just be laughed off the next morning.
>
> *(Craig 2017)*

Such reactions give insight into the intensity of touring and demonstrate the interpersonal effects of its conditions and norms. Feelings or frustrations are withheld to a breaking point, which suggests that crew members struggle to conceal them. This also represents the masculine features of the culture of touring, which will be discussed in greater detail later in this chapter. If there is excessive conflict between particular members of road crews, and to the point that it negatively impacts the tour or production, the offender will usually be removed and replaced. However, research respondents stated that such circumstances are rare. A major reason that such tension occurs is that crew members do not have ample time away from each other nor the privacy to attend to issues and disagreements. Whereas working in a nine-to-five office environment allows people who have disagreements a certain distance, the conditions of touring largely inhibit such opportunities.

> ... you go to work during the day you have a fallout with ... one of your colleagues, you then go home at night and depending on how you see things, there's a fair chance you will regulate yourself out again and by the time you see them again you'll be able to have met some compromise. We don't get the opportunity a lot of the time to have that break from contact and for anywhere as long a time, so if there are any negativities or relationship difficulties or whatever they tend to be magnified because there's no escape from them.
>
> *(Tony 2017)*

This factor further confirms the long-hours culture that characterises working on tour. The working day is not exclusively limited to the hours of a show day, nor to the on-tour persona crew members maintain on the road, but continues on to the forms of transportation as road crew members live and travel together. With the exception of TMs, and generally speaking, crew members' time off, or downtime, is during days when there are no shows and at meal times on show days. However, the catering room is generally occupied by members of the touring party. Spending time apart is essential to managing relationships on tour, but is difficult within the context and conditions of the working lives of road crew members. The next section will explore the role of camaraderie and friendship in road crew members' working lives and the workplace culture of touring.

Camaraderie and Friendship

While it has been established that touring has the potential to cause disruption between people, it also is the source of camaraderie. As in an organised group tour in the context of tourism, the group "reinforces a seemingly independent existence and the ability to co-exist with others" (Schuchat 1983:

469). Mobility tends to foster a sense of closeness (Adey 2017: 10) and can facilitate a bond between crew members. In the closed-off world of touring, people "traveling together and sharing a common culture will feel closely drawn to one another, resulting in a higher level of interaction" (Holloway 1981: 388). Members of road crews often cite camaraderie as one of the reasons they enjoy touring. Tony (2017, 2018) expressed the importance of camaraderie on tour when discussing differences in modes of transportation. Though he now travels by private jet, he was ambivalent about the merits of plane travel in comparison to his previous experience on buses working for other artists. While he admits that plane travel is "very luxurious ... and it's very glamorous," the bus is preferable based on a difference in social interaction: the "level of camaraderie you get on a tour bus tour is pretty high, people who bond fairly tightly." Research on group tourism has found that the back of the bus, or the lounge in a touring context, offers "both location and opportunity" for sociability and that "membership in a tour group encourages this kind of interaction" (Schuchat 1983: 470).

Camaraderie is often expressed through the shared humour that members of road crews use on tour. Humour is a "great unifier" that is "part of the camaraderie that forms from working so closely together" (Douglas 2021: 123, 159). As briefly explored in Chapter 3, workplace humour on tour can be characterised as banter and as deprecating. It is directed at many subjects but often is inspired by the process and experience of touring and the artist for whom crew members currently work.

> It's all kinds of things ... Sometimes, at the artist. Yeah, definitely. Each other. There'll be something that you stick on to for a while and start talking about, like a theme ... Running joke and that will just run and you're just trying to do it to keep each other's spirits up.
>
> *(Joe 2018)*

Crew members make such exchanges in person during the working day or on the bus but may also communicate humour through technology. Joe (2018) explained that his crew has different groups on WhatsApp, one for important work-related information and another reserved for random conversation between crew members. He read a selection of texts aloud from the latter group during an interview.

> ... there's one ... which is just for everyone else talking rubbish. You can turn that off if you like cause sometimes 4:00 in the morning it goes nuts and everybody's just texting nonsense to each other. It's hard to define what it is ... Just jokes, almost like you would do on Facebook sometimes, you know, 'I found this it's funny' ... I mean, our audio technician who went to a wedding, and the band are playing [song

> by artist he works for] and they thought that was funny, so he's just making faces at that you know. Very hard to ...check those two they look like Heart and then ... they're talking about the front row there, they're talking about we should go to the Irish pub across the road, here's our truck crash ... you know, just ... Oh look I've got too many pillows in my room.

His quote highlights the kinds of topics that become the source of jokes and banter between crew members and can be seen to directly reflect key elements of their daily working lives. On the one hand, musicians and the live music environment become subjects of jokes via references to songs and the front row at a concert. On the other, it captures elements of the mobile experience of touring by including comments on the crashed truck, pubs in local areas and the familiarity of hotel rooms. Participant observation gave further insights into the presence and importance of camaraderie and humour on tour. One crew member spent an extensive amount of time making use of art supplies for the sole purpose of teasing a colleague in order to amuse himself. As I have discussed elsewhere (Kielich 2021), visiting touring parties found aspects of the multipurpose work environment to be a consistent source of humour. The use and significance of humour also extends onto the local crew (ibid.). The maintenance of, and expectation to maintain, a light-hearted, humorous work atmosphere illuminates the stress and long hours that comprise it at the same time that it provides a veneer.

Outside of interviews and participant observation, the David Russell Collection archive attests to the significance of humour as an expression of camaraderie and as an effort to maintain morale. Russell, as both a TM and PM, authored daily written communications to his crews that detailed important information about each show day's schedule. The archive contains several folders of "practical jokes" which are essentially documents that look like tour-related communications and seem to have been used for practical purposes, such as to indicate set times. In addition to the key details, he usually included a Polaroid of a band or crew member and some type of commentary that directly made fun of a situation relevant to, or that had recently transpired on, the tour. The source and target of humour was daily life on the road, crew members, days off and the local culture of cities, and the references were sometimes specific to a given artist (Practical Jokes, David Russell Collection). For several years, he also drew a series of comics called "Road Rage," sometimes on the back of tour documents, that directly addressed aspects of the everyday life of live music. It is unclear how they were circulated on the road and who saw them, but their title, topical nature and his very effort in drawing them reveal much about the nature of touring and the daily experience of doing so. Aspects of the content are certainly inside jokes only decipherable by those who were on a given tour, but in general, both the content and the documents themselves offer clear insights into the frustrations of touring and the attempt to overcome them. (Hand drawn comics, "Road Rage," David Russell Collection).

The extensive use of humour effectively implies the difficulties of life on the road. The reliance on it to maintain good working relations and a congenial environment suggests there are underlying reasons to lighten the mood, or "keep each other's spirits up," and a pervasive need to do so. Adrian (2018) described that "it pretty much runs on its sense of humour ... It's kind of like laughing and joking and that gets you through the day really" (Adrian 2018; see also Gorman 1978: 31). David L. Collinson (1988: 185) has shown that humour and banter in the workplace, specifically in a male-dominated setting, is "conditioned by a desire to make the best of the situation and enjoy the company of others." Workers perceive humour as a means to cope with and resist the conditions of the workplace. It provides the "illusion of separation" from the circumstances and strengthens self and group identity (ibid.). Following this, humour may be a feature of the workplace culture that creates bonding and a sense of camaraderie, but it is ultimately a means of coping with the conditions of touring.

The formation of such bonds can inevitably turn into friendships that exist and continue both on and off the road. Research respondents spoke of having met some of their closest friends while on tour. Hawes (2019: 31) stated that strong friendships develop due to "shared trials." Research respondents also addressed the difficulties that friendship can present on the road due to how closeness can complicate boundaries and working dynamics. The challenges of and need to maintain professional relations can be tied to the expectations associated with specific roles. Those in management positions, such as PMs, are responsible for handling aspects of production that directly affect the working lives of technical crews. Though they interact and socialise, they must delicately balance personal aspects in order to maintain the integrity of their positions: "you do need to keep it professional ... cause you're making decisions they don't like" (Joe 2018).

It is in the negotiation of working relationships that the implicit hierarchy between particular crew members is expressed. Tony (2018), who does personal close protection security for a specific band member, articulated that though he finds value in the good nature of their relationship, it can "get really in the way of getting shit done." Tony noted that the level of interaction with musicians has been unique in his career in close protection security, which also included working for diplomats and royalty. In those sectors, which he referred to as the "formal world" of close protection, his "job was not to really interact with my client in any real way, more than perfunctory duties because the relationship can get in the way of the duties." The more informal tone to relations in the music industries can be an asset as it can "help you persuade your client to do stuff that they might not want to do." It has also led to instances where he has had difficulty being taken seriously in his role.

> ... we were in Paris a couple years ago when the attacks happened ... it was my job at the time to run the team and I ordered an evacuation of the building of all of our crew and all of our band, and trying to get the band to get in the car was a real big deal because they didn't understand what was going on and they didn't have that understanding of ... 'if this guy tells me to get in the car in that manner, I better just get in the car and ask questions later.' They wanted to discuss it and eventually I had to say 'fire me tomorrow but you can either get in the car or I am going to put you in the car and that will be it.' Luckily, our guys understand, but the relationship got in the way, because 'oh it's just Tony, Tony can get a little bit harsh at times,' my sense of humour is quite dry and all that sort of stuff.

Friendship between crew members can also be a problem when they are hired based on existing personal relationships. Chapter 2 indicated that working for friends is a common early source of training and experience for crew members. However, as musicians become more successful, close personal relationships may create difficulties if the crew member in question is not highly skilled enough to appropriately manage daily responsibilities on tour. Hiring someone "like a friend or a relative, who lacks the necessary expertise to keep the tour running smoothly" can be "disastrous" (Waddell, Barnet and Berry 2007: 3). TM Malcolm Cook (2011: 15) made the same observation in his autobiography and cited the risk for the artist to take control as being a potential threat to the success of the tour. The need to trade personal ties for professional contacts can be both a result of and assurance for the continued upward mobility of a musician's career. Discrepancies that occur also have the potential to ruin the relationships between the artist and crew member.

These examples combine with findings in this chapter to show that working on tour is marked by tensions in the management of relations. Being a member of a road crew on tour involves different degrees of intensity in the bonds that form between people. At the same time that touring is marked by closeness and camaraderie, relationships on tour can be fleeting and ephemeral. The limited length of a given tour and the uncertainty of any specificity regarding future positions mean that crew members may only work together for a short time. In some cases, the same or similar crews may re-form each time a particular artist goes on tour. In others, it may be a one-time occurrence.

> ... it might be two years if you're part of the core team of a band and you're on the road for two years, then you're gonna spend two years with these people. You might never see them again. Or it might be a month and you

> might never see them again or you might see them again in five years time. So you know that that relationship is transient.
>
> *(Craig 2017)*

In this way, crew members may resist bonding or committing to establishing friendships. The realities of touring that take crew members in separate directions and the potential for emotional implications mean such efforts may not be deemed a worthwhile investment. Similar patterns have been found in a tourism context of group travel. Some tourists refrain from extensive amounts of interaction or "extending their personal lives through the trip" and are not interested in "follow-through," while others seek meaningful relationships (Schuchat 1983: 471). As such, being on tour is at once highly interpersonal at the same time that it is isolating. The chapter will now address the topic of masculinity and how it permeates the culture of touring.

Masculinity, Mobility and the Road

Chapter 1 discussed the male-dominated nature of road crews, which is another example of the widespread underrepresentation of women in the music industries. The notion of "the road" reinforces such marginalisation by its association with masculinity. Rock culture has a strong link to notions of independence and "nomadic bohemianism" (Kearney 2017: 162). Vilanova and Cassidy (2019: 90) identify Kerouac's *On the Road* and the figure of Odysseus as positioning the "archetypical road-traveler" as a man. Furthermore, life on the road is tied to movement away from the realities of everyday life and feminine-associated roles. The "romantic notion" of touring rock musicians represents "a liberation from mundane life and domesticity" (Leonard 2017 [2007]: 60). Historically, men have been mobile, and women have not been unless they have done so in the company of the former (ibid., see Leed 1991). In this way, "travel is 'genderized' and becomes a 'gendering' activity" (ibid.) which is reflected in the norms and expectations on tour. Through their reproduction, gender is a "ritualized performance on tour within a specific set of labor, industrial, and cultural practices" (92). Such gendering can be observed in the language utilised on the road, such as crew members referring specifically to a "brotherhood of roadies" (Douglas 2021: 97).

The masculine coding of touring is further solidified through the mythologising of life on the road in rock music history and culture. While such mythmaking is typically centred around musicians, the associated features can be extended into the lifestyle of touring as a whole. The depiction of touring in the music press and journalistic anthologies describe "boastful tales from musicians of drunken hedonism, sexual adventure, the trashing of hotel rooms and the defenestration of television sets" (Leonard 2017 [2007]: 56). They are also reproduced in music documentaries and films about popular music that glamourise and romanticise touring and the associated indulgent lifestyle (ibid.). Marion Leonard argues that these types of accounts "may be read in gendered terms as narratives of (male) conquest, excess and freedom" (ibid.).

Road crew members are also important figures in this mythologising, both directly and indirectly. Autobiographies written by crew members detail their participation in – and boast about – excessive alcohol consumption, drug intake and sexual exploits (see, for example, Wright 2009). The representation of road crew members is at once part of the myth-making and evidence of their association with these notions of the culture of rock music and touring. For example, the video for the Tenacious D song "Roadie" explores the working life of a newly hired crew member. During his first day of work, he engages in a sex act with a woman who then reveals she wants a backstage pass, proceeds to consume alcohol and inject drugs, destroy part of the dressing room, and finds out he has a child he was unaware of. The substance abuse seems to be connected to rejection and a sense of inferiority and functions as a coping mechanism. In the short-lived Showtime series *Roadies*, crew members are regularly depicted conjuring and participating in over-the-top activities and antics while on tour. Notably, the majority of crew members in these cases are men.

Research findings support these notions and indicate the manner in which masculinism operates on tour. It is a factor that permeates and shapes touring. Masculinity functions as a standard by which various aspects of the workplace norms and culture are set and informed and is reproduced by the working practices and culture of touring. Crew members state that the conditions and expectations of touring lead them to develop a "fairly tough persona" or a "thick skin," to "keep a shell" and "be pretty hard and able to take a lot of shit" (Cooper 2007, Craig 2017). Being on tour necessitates compartmentalisation to successfully adapt to life on the road. Crew members are expected to leave their personal problems behind, handle the conditions and realities of working on tour and do so without comment or objection. Any appearance of weakness is strongly discouraged (see also Douglas 2021: 211). These assumptions about conduct are understood and reproduced by crew members without being explicitly stated.

> … you kind of have to come baggage-free and you have to be able to cope with you know isolation from your family, stresses from travel, the amount of uncertainty that can come along with the job, that you've got to be able to cope with that, you've got to be able to cope with that without complaint. That kind of like strength of character is very important and it's possibly unspoken that that's expected of you. Everybody misses their families, everybody has personal problems, but you're not supposed to

> bring that to work. Whether that's a good thing or whether that's a bad thing I don't know.
>
> *(Adrian 2018)*

These findings address the shared conditions and experiences that come with being on tour and reinforce the need for suitability and adaptability to the actualities of this line of work. Deviance from these expectations is discouraged due to the potential to affect other crew members and the ability for work to be done. "You can't … be in a bad mood one day because that will affect the people around you, you can't be upset one day because you're missing home" (Craig 2017). Crew members perceive emotional difficulties as incompatible with the nature of the work and the pressures of touring.

> You can't have people with emotional problems on the road, it, on a practical level, it endangers the production and on a personal level you know it can start to effect people around you … morale is very important, and trying to carry someone who's obviously in distress, because of the unique working circumstances I think is very very difficult because it's not like an ordinary job and you're not in ordinary circumstances. It does make it difficult.
>
> *(Adrian 2018)*

An inability to cope with the conditions of the road – or the perception of being unable to – can have an impact on the acquisition of future positions.

> … anyone who isn't capable of doing their job … will not get a job again, so any form of weakness, be it physical, mental, inability to go without sleep, well, 'I'm not giving them a job cause he's not capable, he can't take it.' If you can't take it, you don't get a gig, it's as simple as that.
>
> *(Craig 2017)*

These notions are significant enough to the workplace culture of touring that they are reproduced in educational tools designed to inform potential road crew members. The online programme Tour Mgmt 101 explicitly states this in their document entitled "Rules of the Road."[1] The document advises crew members if they "have personal issues, please leave them at home … There's no room for them on the bus," requests that they "do not whine; you are an adult, and hopefully a professional. It's not becoming" and reminds them that "there's no crying in Rock and Roll!!" The fact that such content was written and included as part of the programme implies a need to inform newcomers as well as the importance of adhering to its points, and the title of the document positions the contents as non-negotiable.

These findings further substantiate that crew members acquire positions based on their network but that their ability to be recommended is also linked to how well they fit in with the norms and expectations of touring. The discouragement of weakness or vulnerability is an extension or by-product of the male-dominated and masculine nature of touring. This represents stereotypically masculine behaviour that is collectively shared and reproduced by road crew members. It can also be linked to the exclusionary nature of touring at the level of gender and functions to reinforce those barriers.

Camaraderie is also another means through which the male-dominated nature of touring is evident, as it leads to a tendency for stereotypical masculine behaviour to permeate the workplace culture. Male research respondents acknowledged that all-male road crews can engage in stereotypically "macho culture." They characterised it as "blokey" and as comprising "male banter" such as "dick and bum and fart jokes" (Craig 2017, Adrian 2018, Joe 2018; see also Gorman 1978: 33). They also expressed a personal dislike of macho culture, seeing it as offensive, unsophisticated and unbalanced. The maturity of the respondents and the number of years they have encountered such behaviour during their careers may be contributing factors to their perspectives. Male respondents stated that they welcome women on the road, cited them as positive influences on the culture of touring and indicated that they believed women's presence worked to undo or balance the atmosphere.

> I think the gender balance of the crew is very important in that sense in that if it's all male it can be quite, pretty horrible. And I think having a number of women on the bus can change that dynamic quite dramatically and it becomes a little bit more nurturing is probably a bit too strong a word, but you don't get that whole big macho loads of blokes on the road together thing. Cause a lot of the women'll just cut through that bullshit. And that stops all that a little bit, it changes the dynamic and it makes it much better.
>
> *(Craig 2017)*

Members of road crews take issue with and become weary of macho culture. The acceptance of women on tour is ultimately a positive indicator and suggests the potential for a more inclusive workplace. However, the reasons for their welcome presence, and the manner in which it is expressed, effectively reduce women to their gender and stereotypes about their roles. Women are welcome on the road for the ways in which they can create a better atmosphere and make touring more comfortable for men. This effectively reproduces male dominance and stereotypical gender roles.

A masculine culture contributes to challenges for women on tour and shapes their everyday working lives. Women recount a range of experiences related to being minorities on tour. They are subject to derogatory language (Hawes 2019: 341–342), report garnering respect from male counterparts on the basis that their presence in the backstage area did not revolve around sexual pursuits (31) and likewise enjoying relationships with male colleagues that was not sexual in nature. Women also attest to feeling supported and protected by male crew members (342, Douglas 2021: 46).

At the same time, research respondents spoke to a lack of tolerance for poor treatment of women on tour. The former two accounts were from crew members whose careers began in the 1970s, and this may be indicative of gradual change. Adrian (2018) stated that women are respected and that he "think[s] anything like that would be stepped on fairly quickly. Certainly from the people that I know, that wouldn't be tolerated."

Interaction between crew members is only one site where the gender imbalance can be felt. Audiences are also complicit in the upholding of this masculinist culture as demonstrated by their reactions to the sight of women road crew members. Douglas described the aspect of her job that she found the "most uncomfortable" in her early career was having to go onstage in front of an audience (66).

> Eventually, they would figure out I was a girl and start whooping and hollering; it was always embarrassing to me. During this second set-up, a bored, restless and drunk audience did just that. 'Look, it's a Sheila! One heckler called, while another thought he was funny with, 'Don't hurt yourself, little girl.' I hated it but just got on with the job, pretending I couldn't hear any of it, hoping it would be over soon.
>
> *(66)*

The challenges associated with being a woman on the road were not limited to male audience members. Douglas also encountered difficulties with female fans of the band AC/DC, who perceived her as a threat due to being in "close contact with the band" and would regularly call her "'bitch' and worse" (90–91).

In these ways, women encounter various tensions, have both positive and negative experiences and confront challenges in their working lives, which is largely connected to their status as minorities and the masculine culture that permeates the culture of touring. The chapter now turns to explore the feelings of disorientation that touring creates for crew members.

Disorientation

For musicians and road crew members, "time and location get blurred when you're moving so quickly around the globe" (Barker 2017) and "sometimes you don't know what city you're in" (Simpson 2015). The possibility for touring to create a sense of disorientation is among the clichés and stereotypes of life on the road. It has most commonly been addressed, with a degree of humour, in relation to musicians when they are performing and forget which city they are in, which has been described as "one of your worst nightmares" (Barker 2017). The "rockumentary" *This is Spinal Tap* (Reiner 1984) is a common reference point used to characterise this experience. The film features a well-known scene that takes place on a show day in Cleveland, during which the band members are lost in the backstage area and one of them repeatedly states "Hello Cleveland!" in an effort to commit the correct city to memory before walking on stage. Members of road crews are also subject to such experiences. As example, in February 2017 at a concert in Australia, Guns N' Roses' long-time instrument technician, McBob, introduced the band to the crowd by referring to the audience as "Sydney" and did so by screaming the city's name (see Frankel 2017). However, they were actually in Melbourne, nearly 900 km away, and the situation was made worse by the fact that McBob is actually from Melbourne.

My research findings support the notion that touring can be disorienting for road crews, and that such an experience is part of everyday life on tour. A sense of disorientation was repeatedly observed during participant observation. Informal conversations with members of road crews were often held while walking and directing them through the venue. These talks typically included a question about where they had just come from. In all such instances except one, the crew member could not immediately remember the previous city. Their response usually involved a pause or hesitation, and they often consulted their mobile phone to check the itinerary. In one case, a bus driver could not remember where he had arrived from despite being the person who drove. The only crew member who could recall the previous location without hesitation was a younger, female assistant TM, which could imply the degree to which she was still trying to prove herself. In another instance, as a visibly exhausted band member talked with crew members, they referred to their previous concert as having occurred a couple of days before, when it had actually taken place the night before.

The sense of disorientation is linked to a tour's itinerary. An itinerary is essentially the plan for the tour and comprises the tour dates, individual show schedules, travel agenda and accommodations. The itinerary is represented by the tour book, which is a physical document that includes specific details of the tour plan, in addition to the names, job titles and contact information for involved companies and personnel.[2] Though itinerary information is communicated via email and smartphones (see Chapter 5), it is still customary to print the tour book (Reynolds 2008: 358). Itineraries are also highly subject to change, and crew members colloquially refer to the tour book as the "Book of Lies" because "as soon as you print an itinerary it's out of date" (Joe 2019).

The primary influence on an itinerary is economics. Touring is expensive,[3] and artists need to play as often as possible to offset costs. For example, in André Novoa's (2012: 353–354) study of grassroots musicians on tour in Europe,

Example 1		Example 2	
23 May	New Orleans, LA	10 July	Brooklyn, NY
27 May	Raleigh, NC	11 July	Charlottesville, VA
30 May	Greenville, SC	12 July	Philadelphia, PA
1 June	Lexington, KY	13 July	Boston, MA
3 June	Fort Wayne, IN	15 July	Montreal, QC
6 June	Madison, WI	16 July	Toronto, ON
8 June	Green Bay, WI	18 July	Pittsburgh, PA
11 June	Moline, AL	19 July	Cleveland, OH
14 June	Arlington, TX	20 July	Chicago, IL
22 June	San Diego, CA	21 July	Royal Oak, MI
26 June	Phoenix, AZ	23 July	Minneapolis, MN
28 June	Las Vegas, NV	25 July	Calgary, AB
29 June	Las Vegas, NV		
6 July	Vancouver, BC		
10 July	San Jose, CA		
13 July	Los Angeles, CA		

FIGURE 4.1 Schedule of tour dates, two examples.

the group's schedule included 19 days of travel and only one night without a show. In contrast, musicians who have greater available resources and highly lucrative tours, such as Paul McCartney or The Rolling Stones, are able to take two or three days off in between show days. Figure 4.1 features two examples of schedules of tour dates, the first from a top-tier artist and the second from a mid-level artist.

Figure 4.1 highlights the additional, and related, importance of the routing of a tour, which is the relationship between where the concerts take place and how the touring party travels to and from them. Difficult routing can increase expenses and negatively impact the health and wellbeing of the touring party. Members of road crews use the term "dart board tour" to refer to instances when the routing is so complex that it suggests it was "determined by throwing darts at a map" (Thomas 2020). The example used was a four-day stretch of a tour that went from Montreal to San Francisco to Atlanta to Pittsburgh. These destinations were clearly chosen for effect, but it implies the importance of routing. For road crews, these factors are significant because they affect the character of their working lives. The itinerary dictates the schedule that they will follow and therefore shapes the experience of touring. It is interesting to note that the pace and related experience of concert touring are not unlike that of organised tourism. The belief that tour guides "need to pack every moment with activity" leads to travellers becoming fatigued and disinterested partway through the tour (Schuchat 1983: 472). While such practices could arguably be more easily adjusted in a leisure-based setting, the underpinning concern of economic value and the resultant pace affects the intensity in much the same way.

Research respondents described a variety of occurrences that suggest the potential for touring to be a source of disorientation. Though crew members may have some engagement with the geographical locations they visit, travelling on tour is practical and functional, and its overall purpose is to facilitate the realisation of individual concert events. Travelling on tour can effectively be defined as the process of following the various details of a tour's itinerary. Road crew members' consistent movement between different versions of the same types of spaces can inhibit differentiation from one location to the next.

> So yeah, that's the one thing that people always ask you, must be amazing travelling. No, I've seen the airport, I've seen the Holiday Inn and I've seen a big tin shed. And I've looked out the window of the car between the three. You might get to go for a half-hour walk around about lunch time.
>
> *(Joe 2017)*[4]

Disorientation also has the potential to arise when crew members encounter similar itineraries on different tours.

> … two bands I mainly work for now. And it just so happened that last year it would pretty much be one band would finish touring and I would go straight onto the other one. I spent a lot of time in America, I started on one side, Miami with one band, went across to LA and then you know, I … then picked up the other band and we came all the way back playing some of the same towns I'd been in not long before. There's a lot of repetition goes on, too cause a lot of bands play the same venues. You know there's certain venues on the circuit and then depending on your size you will sort of go up or down depending on how you're doing.
>
> *(Duncan 2018)*

An itinerary does not necessarily entail a full tour schedule, but can also include one-off dates. Craig (2017), based in the UK, described having once "flown to Australia for one show and then flown back again." In such a case, disorientation can be linked to the intensity of travel and drastic changes in time zones rather than the continuous pace of moving locations during a tour. The manner in which touring parties travel can also be a contributing factor. Tony (2017) works for an artist who has their own 737 and, because he "directly serve[s] one of the band members," is among the 54 people who travel on the plane. Rather than fly from one city to the next, the travel agenda is based on a practice called "hubbing."

> … if we go to the states, they won't fly from New York to Detroit to Texas to whatever, they will split the states in half and they will hub out of either New York or LA, so instead of doing what the crew do where they go on the bus in San Diego and they travel all the way up the coast in a bus to San

> Francisco, we will stay in LA, we will get up in the morning, we will get on the plane like it's a bus, we will fly to San Diego, we will do the show, we'll fly back to LA, we'll all go to bed at the same hotel. The next time we'll do it again and that's how we do it you know. And you have certain rules where if the flight is over, I think the rule is if we're over a four-hour flight then we may stay over where we are just to save flying back four hours, but yeah, we tend to hub it and that goes for Europe as well.

For crew members like Tony who work for artists who primarily perform in arenas, the characteristics of such types of venues can also contribute to a sense of disorientation. Road crew members and musicians alike have described North American arenas in particular for how they "seemed identical," "felt like you could be anywhere" and are "sort of all the same but with just enough differences to mess with your head and keep you slightly confused" (Edgar et al. 2015: 136). Arenas can be linked to disorientation based on their generic appearance and because they are typically located in similarly indistinguishable areas in the outskirts of cities.

Disorientation on tour can also be understood from a different perspective. On tour, the working lives of crew members, and their spatial and temporal whereabouts, are structured according to the itinerary and demands of their roles. In this way, they are not so much disoriented as they are exhibiting a lack of concern and attention. They are oriented to the activities and places of work rather than the wider location around them. Given that the daily objective on tour is the realisation of live music, the specific location in which a road crew works on a given day is ultimately irrelevant. In the same way that travellers on a guided group tour in the tourism industry exist within their own "bubble" and are only acclimatised to a local area through its limited framing by the tour guide (Holloway 1981: 382), road crew members are primarily oriented to an area based on the work-related spaces and details of a tour's itinerary. An individual stop on a tour is one in a series in which crew members attend to the activities involved to ensure a concert happens. The location serves as a temporary setting that road crews occupy and is a container in which the necessary tasks can be enacted and completed. The ability to do their jobs and the end result are more important than the details of where they do it, and their focus is on the work and the deadline rather than the location. In other words, the fact that it is a "show day" is more important than the fact that it is occurring in "Chicago."

This experience can be characterised by the difference between "location" and "place." The former can be a city, or thought of as point A or point B, but it does not have meaning. On the contrary, the latter are "locations imbued with meaning and power" (Cresswell 2006: 3). Similar to Felski's (1999: 24–25) notion of "home," they are "meaningful segments of space" to which people become attached, fight over, exclude others and experience (ibid.). For road crew members, the various buses, hotels and venues they temporarily occupy and in which work and life are experienced while on tour function as "place." Road crew members remember and speak of particular places based on the work conditions they present. Venue A was distinguishable because of the distance between the load-in area and the stage, and more so once it had been modified to be more accommodating. Boston is particular for how the load-in at the Orpheum Theatre is through the front door due to the fact that there is no back entrance, and Glasgow is associated with the three stories of stairs that are a feature of the load-in at Barrowland. These factors suggest that crew members are oriented to an area based on the features that relate to the particular context of their working practices. In this way, disorientation regarding location is the result of an orientation towards tasks and place. The chapter now examines the potential for touring to affect the mental and physical health of road crew members.

Mental and Physical Health

The conditions in which road crew members work can be linked to the possibility of adverse effects. Research findings offer evidence of the need for the concern and scrutiny over the working conditions in the music industries in particular relation to mental health. One factor is the aforementioned long-hours culture in which road crews work, and which is important for understanding the everyday life of live music. Mark Banks (2007: 56) has noted that cultural workers are valued because they "show dedication by working long hours or [working] under oppressive circumstances." To be "described as a regular '9–5' worker is no longer a commendation of diligence but a term of disapprobation" (Banks 2007: 56). In other words, the grit and endurance of such hours and conditions are respected and potentially rewarded. In 2020, The Tour Health Initiative was launched, which is designed to collect data from members of road crews on the "conditions and effects of living our lives on the road" (tourhealth.org). The survey includes a question about working hours but is still in the data collection stage and conclusive findings are not yet available.[5] This topic has also been recently addressed in a report by the Broadcasting, Entertainment, Communications and Theatre Union (Bectu)[6] that found long hours to be the biggest concern of cultural workers (Evans and Green 2017: 3, 5). The report observed that people who work long hours have a poor life-work balance and that

> the long-hours culture is a real danger for workers. It means that people suffer mental health problems, anxiety attacks, disrupted sleep patterns and poor emotional well-being — and strokes, heart disease and even cancer. People make mistakes that affect the quality of their work.
>
> *(7)*

Though specific to the long-hours culture of the film and television industries, the similarity in working conditions suggests that these conditions are widespread throughout cultural sectors. Evidence of these patterns amongst road crew members was found during research for this study and is detailed throughout this book. The authors of the Bectu study referred to the long hours as an "unnecessary" component of film and television production. However, as suggested by respondents in this chapter, it is taken as a norm in the contemporary cultural workplace. Working "longer or unsocial hours" (Banks 2007: 56) is among the attributes that Angela McRobbie (2002a, 2002b) argues are characteristic of "being flexible" in the cultural workplace, which means that "one must do whatever is required to support commercial interests." In this way, the long-hours culture that characterises working on tour illuminates the great potential for road crew members to be exploited and exposed to risks that impact their health and wellbeing.

Current media coverage and studies conducted by UK-based organisations such as Help Musicians UK,[7] Music Manager's Forum[8] and Music Support,[9] along with Sally Anne Gross and George Musgrave's (2020) book *Can Music Make You Sick?*, also report the potential for working in the creative industries to impact mental health and wellbeing. The publication of the 600-page book *Touring and Mental Health: The Music Industry Manual* (2023) also attests to the significance of this issue and the need for guidance. Attending to mental health is seen as a priority as touring in particular was cited as "an issue for 71% of respondents" surveyed (Britton 2015, George 2017, Gross and Musgrave 2020). Though the focus is primarily on musicians, the shared conditions of touring suggest that a comparable situation exists for road crews. The Tour Health Initiative survey aimed at road crew members includes questions about mental health, though, as previously mentioned, results are as yet unavailable. Generally speaking, attitudes towards mental health in live music have begun to shift, and efforts have been made to encourage openness about discussing related issues and eliminating the stigma around doing so. Organisations such as Tour Support[10] promote awareness and develop platforms for addressing these issues on tour. The break in touring due to the COVID-19 pandemic was viewed by many road crew members as a time to rethink and revise current practices.

In the particular context of this chapter, mental health is a consideration in relation to the close living and working conditions of touring and likewise the kinds of support crew members may receive from each other. Attitudes about weakness and the impression management that crew members rely on suggest that dialogue about such issues is minimal, and research respondents reported different experiences concerning the degree of openness. While respondents acknowledged that the norms and expectations have certainly inhibited crew members from asking for help or talking about difficulties during the course of their careers, they also state that such conversations have become more commonplace than in the past. They indicated that when crew members encounter difficulties, they often do confide, though trust is an essential component. As this chapter discussed, the bonding potential of touring means friendships can and do develop and that crew members may find others among them who they feel they can "talk to about stuff" (Craig 2017). If members of road crews are having a difficult time, they will let others know. Experience is an asset in such circumstances based on an understanding of the pressures that touring can create.

> … if somebody's struggling with something people aren't usually shy about letting people know … I properly barked at somebody and then went to apologise and said, 'I'm just under a bit of' and they understood. It's just you do trust each other to a certain point, and everybody's who's been doing it for a while understands that it's really difficult to do for a long term without some sort of pressure release here and there. So you do offer an ear or a shoulder or a drinking buddy or a walk in the country, you do try and look after each other, or people make unlikely friendships you know, whoever suits you most you'll end up with, hanging out with, some people don't talk much, but you know, it's just them.
>
> *(Joe 2018)*

Crew members also offer mutual support when dealing with life events. As touring involves extensive periods of travelling, they experience life changes and encounter personal challenges while on tour, which their colleagues witness.

> … you see people going through divorcing. You know you see people have children while they're away, you've seen people trying to give up smoking. I'm not smoking on this tour, oh fuck, here we go, but you see people lay completely bare.
>
> *(Andy 2017)*

> I think you'll find that the whole tour bubble look after each other … It's natural thing people on the road look after each other. Towards the end of the last tour there in North America my mum was really sick and everybody was asking after her, they all said, the management said, 'if you need to go home, just get on a plane.' And a number of them sought me out to talk about how I was doing to talk about similar situations with them, I missed the last show in London cause my mum died that morning.
>
> *(Joe 2018)*

The shared experience of going through major life events at a distance while doing so in close proximity with others creates a type of bonding through relatability that encourages the formation of a support system.

Specific roles on road crews foster care-based working relationships that are ultimately unbalanced and unreciprocated. The nature of these roles means that the people occupying them tend to become caregivers while their own needs may go overlooked, disregarded or neglected. Tour management involves "dealing with other people's problems," and people "don't necessarily appreciate, or even seem to entertain the possibility, that you have problems of your own" (Hawes 2019: 339). The responsibilities of tour management, as will be described in detail in Chapter 5, and the authority granted the TM, directly reinforce this dynamic. TMs are "responsible for people, looking out for them, making sure they get what they need" (ibid.). The working relationship between TMs, other crew and musicians is therefore often unidirectional; TMs take an interest – or must do so on account of their roles – in other people who may not, or may not have to, reciprocate (see 340). This means that, as a by-product of job responsibilities, there is a particular effort made towards the wellbeing of others by one individual.

As we have seen, particular roles on tour come with gendered associations, and this factor can have effects in relation to mental health. Tasha's (2018) administrative roles also place her in the role of confidante. As the person who assists with the everyday requirements for the crew, she is also often the default person for crew members when they need to talk to someone.

> I don't think people do talk about it very often … sometimes I'm in a position where people come and talk to me if they're having a bad day just because I'm their 'mom' or whatever on tour so people come and tell me, I might hear it more than other people do.

These types of roles are the positions on tour that are commonly occupied by women. These roles can have similar features to the more personal side of tour management (see Chapter 5). A crew member who is both a woman and working in a "mom" role is deemed safe for discussing problems. In other words, expressing "weakness" is less threatening and more acceptable when it is done in the presence of a perceived caretaker or someone occupying a maternal role. Though Tasha is available to offer support to others, she also felt as though she had to conceal her own emotions "because I'm the one that's there to try and make things good for everyone, I would have to hide my feelings. I would try." Tasha perceives herself as responsible for the wellbeing of others and that her role requires her to do so based on its expectations and gendered associations. Following from this, she refrained from showing her own feelings, which can be linked to the social norms of touring as well as the gendered expectation to put others first. It is also connected to her status as support personnel and demonstrates a larger point about how road crew members must often put others before themselves in their working lives. This case, however, also reveals that this can operate along gendered lines.

Particular roles on tour may be more vulnerable than others based on the nature of their responsibilities.

Tasha spoke to becoming "burnt out" from having to conceal her feelings during a particularly stressful tour. This example further suggests that such expectations can place unrealistic demands on road crew members in terms of wellbeing and have the potential to negatively affect them. She recounted a "toxic" tour during which a crew member who was dating the artist manager used his relative power to have several other crew members fired baselessly. In this situation, Tasha's specific role again made her the person everyone voiced their concerns to.

> ... people were upset and so like I was hearing about it from everyone, but I had no one I could vent to about it, like how stressful it was for me and how terrible it was for me. Cause I'm seeing my friends getting fired or quitting and that was really hard for me, and I had no one to vent to about it except for my friends back home ... [who] were sick of hearing about it and were like 'why don't you just quit cause this isn't good for you?' And I wouldn't, I couldn't quit. It's hard to complain to someone who doesn't work on the road because if I talk about the everyday little things that happen, it's just they add up to a bunch of big problems.

Tasha and the other crew members likened the tour to a war and said that the environment was such that people who "haven't been there with that particular artist" cannot fully understand the extent of the situation. This illustrates how much a tour can vary according to the artist a crew member works for and the extent to which musicians can affect their daily working lives. One of the ways that crew members coped was to form "suicide pacts." Particular crew members agreed that one of them "won't quit unless you quit kind of thing, or if you get fired then I'll quit." After the tour, she went to therapy and took time off from touring in order to address and recover from the effects of the situation.

Amy (2018) discussed a stressful tour and said that she ultimately decided to stay in order to ensure she was paid. As a favour, she took over for an artist who had recently fired their previous TM. The crew was "great," but the artist was a "challenge."

> ... after that first run, [I] really did not want to do the rest cause I was like, I knew it, that this was way too hard on my mental health, and I thought to myself, 'you know, I could eek out two more weeks because if I walk away somebody else is going to get paid for all the work I've done and put together' and ... I knew that this tour was no good for my mental health and my anxiety levels, it was really there was just a lot of pressure I was being

> put under and I knew that for me I should just step away and then I thought this guy is never gonna pay me if I just step away and then I've done all this work, so I'm just gonna power through for two weeks.

Despite being hard on her mental health, the advance work she had done as TM meant she would lose the money should she choose not to finish the tour. Such circumstances highlight the precarious aspects of working on tour. The absence of job security puts road crew members in a position in which their wellbeing may be compromised. They not only recognise the adverse conditions in which they are working but also understand the pressures of their line of work and must prioritise their livelihood. They effectively choose between their work or wellbeing. Tasha and Amy chose relatively extreme language to characterise their work environments and experiences within them. The tour was likened to a war and working meant having to "power through." Such language implies that these crew members encountered strain and that the tour required considerable effort to endure. Though these circumstances represent particularly challenging tours, they draw attention to tensions that exist in the working lives of road crew members and potential sources of mental health issues on the road.

The culture and conditions of touring can also have adverse effects on the physical health of road crew members due to the expectations and pressures that the constraints of the itinerary impose on them. Craig (2017) stated that road crew members "can't get ill" because the schedule of the tour essentially inhibits room for sick days, and no one else is available to temporarily fill the role. Similar to musicians, crew members do not have replacements readily available should someone become unable to do their job. Working on tour is inherently precarious as it is constituted by conditions that presume and rely on the good health of all members of the touring party without means of accommodation for situations that occur. The result is that if road crew members "do get ill you just have to do your job, even though you're ill" (Craig 2017). As a result, crew members may neglect themselves or delay receiving attention to problems they are having.

Tasha (2018) recounted having a broken foot on tour and receiving minimal assistance from the other crew members. She stated that everyone was aware of her injury and helped her for the first day, but that they "soon forget, you know. And it's not saying that they're bad people, it's just how it is." Her experience confirms a relative lack of tolerance for any type of weakness and shows the difficulties of handling physical limitations on the road. Additionally, the choice to tour with an injury or illness can be linked to the freelance nature of this line of work and to the importance of staying visible on the circuit discussed in Chapter 1. These findings also imply that the conditions of touring are accepted and taken for granted by crew members, and represent a denial of the potential difficulties and an inability to change them.

In other cases, a health condition may force crew members to leave a tour. Amy (2018) recounted having shingles while being on tour. Dealing with the infection meant that she was "burning the candle at both ends" and needed to "get this under control." In October she knew she was going to have to leave the tour due to the condition but stayed on till mid-December to finish the first part of the tour. Someone replaced her for the remainder of the tour that began the next year. Though she did ultimately leave, she still waited before attending to her condition, which further illuminates the pressures road crew members encounter, the precarious nature of working on tour and the tensions that exist in their working lives. Kim Hawes (2019: 171), following extensive touring with Motörhead, recounted trying on her wedding dress six days before the ceremony and discovered that her weight had shifted from a size 10 to a size 8, which she attributed to "the physical stress of touring." The final section will analyse the experiences road crew members have when they return home.

Off the Road

Being off the road is also a significant component in the everyday lives of workers in live music. While tourists may express relief at returning to the comforts of home after a trip, many road crew members experience a sense of disruption when leaving a tour. Returning home offers less a sense of relief about going back to somewhere familiar and more of an adjustment process of breaking with and transitioning from the patterns and activities of being on tour. It represents a marked contrast in daily rhythms and responsibilities, and the change affects crew members in multiple ways. My findings suggest that touring can function as a stabilising feature in their lives and that their identities are strongly shaped and defined by touring. That touring can create a sense of normalcy for crew members is evident in the amount of time they spend working versus being at home and the adjustment they encounter. During participant observation, a crew member stated that he had been home for approximately fifteen days that year and, due to the infrequency with which he sees his wife, still uses terminology reminiscent of newlyweds despite being married for more than 20 years. As Felski (1999: 28) observes in her comments on everyday life, and which can be applied to touring, if identity formation results from repeated behaviours and emotions, any sudden change in the "rhythm of one's personal routines … can be a source of profound disorientation and distress." The disruption felt when workers return home from touring shows its significance to their identity.

When returning home from touring, road crew members must readjust to the norms of life off the road. Returning home means that they abruptly stop working, travelling and following an itinerary that they have become accustomed to throughout the duration of the tour. They pause from the constraints of time and demands of a schedule. The long hours, unexpected problems and deadlines cease to dictate their lives. In this way, they must adjust to a pace of life that differs from the pace and pressures of an itinerary. Road crew members manage the residual effects of life on the road as they re-enter the patterns of daily life at home. Doing so is a process, and the experience is effectively an extension of the long-hours culture of touring.

Leaving a tour reveals the disconnect that crew members perceive between being on and off the road. It further substantiates that they experience touring as another "world." Tasha (2018) likened the experience of coming home to "re-entering earth from outer space." She identified the difficulties in transition as being related to aspects of the norms and culture of touring in distinction to the modes of conduct and responsibilities associated with being at home.

> ... cause you really are sort of in this weird bubble and you have to get your legs again underneath you ... I come home from tour and I need, I need a good two days to just sleep and sort of get, like stop swearing like a sailor and you know get back in the habit of cooking for myself and ... doing self-care that gets put on hold when I'm on the road because I'm caring for everyone else.

Her emphasis on attending to basic needs and to taking care of herself suggests the extent to which touring compromises the ability to do so. Furthermore, her role often involves putting others before herself. The need to readjust to cooking is a response to the convenience of catering and reliance on purchasing food on tour, which does not require crew members to prepare anything for themselves, and, in any case, renders that impossible most of the time given the available amenities. Joe (2017) also recounted difficulties with acclimating to the norms of everyday life off the road. The regularity of showering in hotel rooms on tour caused him to often forget to bring a towel into the bathroom at home. The other difficult adjustment was the free and frequent access to alcohol on tour made available by the hospitality rider (see Chapter 5).

> ... I still have trouble sleeping without a glass of wine because of my habit of you finish work, you get on the bus, you have a quick couple of glasses of wine, you go straight to sleep cause you need to get that six or seven hours of sleep on the bus in order to be up and function the next day. And wine for me then became a trigger. Drink wine fall asleep. Don't drink wine don't fall asleep ... I've still got it. You can call it minor alcoholism. But I need it, I rely on it and I've yet to manage to stop it.

The time constraints of the itinerary created habits on tour that led to a physical association between alcohol and rest and continued into his daily life when he returned home. Craig (2017) similarly articulated that the nature of the lifestyle of touring becomes more apparent when crew members stop and realise they had engaged, without being aware of it at the time, in unhealthy behaviours which can include lack of sleep, irregular eating, alcohol and drugs. He recounted the practice of taking drugs in order to stay awake for longer when he was tired and still had work to do. He explained that "you learn very quickly that that actually is counterproductive" but seems like a good idea at the time given the pressures and circumstances. This suggests that aspects of the workplace culture are taken for granted by crew members and that the pressurised environment of touring normalises and encourages such activities. This shows the impact that the conditions of touring can have on crew members, how they cope with them and how they carry over. The conditions of touring do not solely affect crew members while on the road but continue into their non-touring lives and lead to longer term habits or practices.

Research respondents reported a variety of sensations when returning from tour. Tony (2017) described that for "the first three days you're bouncing off the walls." Adrian (2018) likewise needed a certain amount of time to adjust due to the sudden shift and agitation that follows leaving a tour. He linked this directly to the difference in schedule and routine.

> Mentally, I'm a lot better than I used to be … I'd have what I used to call a restless period for a few days, cause after a long time where you've been working to a certain schedule which … usually involves finishing late at night, then getting back into a different [one], it can be like a horrible grinding of gears changing your routine overnight. So I used to have like this come-down period of a few days while my body recovered and I got into the rhythm of a different routine, a non-touring routine, but that changed when I had children. Cause as soon as you come through the door you're on duty, so that got rid of that habit fairly quickly.

He highlights that experience played a part in making the adjustment easier, which suggests that crew members learn to manage the conditions of touring over time and make efforts to adapt. Adrian found the transition is eased by his family due to the pressure it placed on him to shift from his role on tour to being a father. His reference to the come-down period as a "habit" and that having children put an end to that practice suggests that such restlessness can be avoided if other responsibilities fill the place of touring. Andy (2017) also attested to having an easier adjustment than when he began touring, which he linked to maturity. Rather than children making it more manageable, he used to find coming home to his family "extremely difficult" due to how domestic responsibility was at odds with the "sex, drugs and rock and roll mode" that he would find himself in.

Throughout his career, Joe (2018) found that he was "either on, a thousand miles an hour, or off" when he returned from tours. He would try "to organise everything" at home and come up with a series of household projects or renovations to work on to occupy himself. At the latter extreme, he "become this sloth. I did nothing but stick my head in the fridge and sit on the couch." The low that occurs following a tour is called "post-tour depression" or the "post-tour blues" and has been experienced by both road crews and musicians. Joe never experienced depression before he began touring regularly and for extended periods of time. He connects it to the ways that the pressurised work environment of touring and the related heightened state of alertness and continuously being "on" are in marked contrast to the pace of domestic life when the urgency of an itinerary is removed.

> ... when you're on the road you have no choice, load-in is this morning, the show is this afternoon, flight is tomorrow, all these things need done. And if they don't get done, you don't get the plane and the whole thing comes crashing down.

Crew members find the fast-paced and deadline-driven nature of touring to be a sharp contrast to the norms of everyday life off the road. Kim Hawes (2019: 340) felt "shock" at how "normal life takes forever." She recalled being particularly struck by the shift between the demands and effects of a pressurised environment such as touring and everyday life. As a TM, she explained, she became accustomed to a sense of immediacy and to things being taken care of right away. Being off the road and having to a wait a week for a repair or other service to be attended to seemed "a lifetime away" (ibid.). While TMs are used to being the ones being depended upon, Hawes also realised she no longer had a team to rely on when she left the road and experience a sense of having to "deal with obstacles I hadn't even noticed before" (ibid.). Without having the rest of the crew to assist with whatever it was she could not do herself, she had to do it herself.

As a TM, work does continue at home in the form of advancing the next tour, but there may be a period of time before such preparations are necessary, and the pace of paperwork, phone calls and emails is markedly different. Joe found difficulty motivating himself to start working again after coming home.

> Previously I'd get home and I'd be a mess for a week. And then I would try and do some work and I couldn't face it and then I wouldn't face it and then it would be late, and then I'd be guilty and then I'd just develop this whole difficulty in facing up to it in between. And as soon as I'd go back out on the road I was fine.

The after-effects of touring lead to a lack of motivation to work, which can in turn have implications for tasks being completed. Such difficulties are alleviated, however, by going back on tour, which creates a sense of normalcy and is a stabilising feature. Joe eventually recognised the need to create balance when returning home by maintaining some type of consistency. Doing so is assisted by learning to minimise the difference between work and life and "live on the road as you would at home and then learn to live at home as you would on the road." Continuing to be busy, establishing a routine and resisting the temptation to lounge are important components.

> ... the balance is I need to keep getting up early, I need to keep getting up at half six to go to the gym now ... I just need to be slightly active in the morning ... Get up, start doing things. And now I've taken a desk and an office around the corner cause working in my place is just really bad for my head. I can't do it anymore. So I go for a swim, cycle in, got my desk, do three-four hours a day tops. Then I'll go back and make dinner and paint the shed and [his daughter] will come home from school and we'll do that stuff. But as long as I'm a bit active for the first half of the day I stave off depression.

Craig (2017) expressed a similar experience of having learned to establish consistency between life on the road and home. The lifestyle of touring, with access to and consumption of alcohol, the quantity and variety of foods at catering and while travelling and the relative lack of sleep, can also factor into the post-tour experience. He became dissatisfied with how he would feel when returning home and made changes to develop a more consistent and healthier way of life.

> ... I think a lot of people quite often would get ill when they came home from tour, that's something that is a consequence of the life that you lead on the road, and I got really fed up with that and I made adjustments to life on the road to make it healthier so that that didn't happen because it's really frustrating, you get home and you're suddenly in bed for two days and then you go back on tour again ... So that I would be in a fit state when I got home, not be a wreck when I got home. Sometimes it's not possible, you just haven't slept for a week ... But if you can make it as healthy as possible so you get home in a good state ...

The manner in which crew members live on tour can have continued effects and directly shape their lives at home. Touring therefore shapes the lives of road crew members on the road as well as off. Particular aspects of touring, such as the constraint of the itinerary and pace of everyday life, have effects that are beyond the control of crew members. At the same time, factors related to lifestyle can ease the transition between home and the road.

Not all crew members encounter difficulties with adjusting to home life after being on tour. Duncan (2018) stated that he is "well adjusted" to coming off the road and able to "switch off into being-here mode really." He does not find a major difference in terms of "head space" between being at home or at work, which suggests he does not experience the same agitation or depression that others may. The only problem he encounters is being tired, and he likened it to the fact that crew members "don't really have weekends" on tour, so he compensates when he returns home. Rather than the restlessness and depression described by other research respondents during their immediate return, his first few days involve resting. Duncan stated he has never experienced post-tour depression and is normally happy to come home. Amy (2018) described a similar pattern when she returned from tours. During her career, her experience of post-tour depression shifted from more intense to minimal, and her transition period involved sleeping, getting organised and running errands. The change was based on becoming "less emotionally invested" in touring than during her early tours. In the past, the camaraderie of touring was more important, but over time the recognition that it is a job and not a social event, combined with moving into tour management and business-related aspects, led to greater emotional detachment. From these findings, it can be said that returning home from tours affects road crew members in diverse ways and that their ability to adjust may change as they gain more touring experience.

The tasks and patterns that structure the lives of road crew members on tour can come into conflict when returning to their homes and families and present difficulties in their interpersonal relationships. Part of the challenge is to switch from the role on tour into the role in relation to the family or partner. When crew members leave home for a tour, they leave behind a partner or family who subsequently adapts their daily lives in the former's absence. The extent of the crew members' involvement with the daily activities of life back home during a tour is through phone or Skype calls from the road. Absence and compartmentalisation can have effects on both the crew member and the significant people in their lives.

> ... it can actually make you a little bit less sensitive, it can de-sensitise you to normal life to a certain extent, so when you get a phone call telling you that your boiler's on the blink at home, it doesn't mean anything to you when you're on the road because ... it's not in that sphere of, I don't know, influence if you like.
>
> *(Tony 2018)*

Tana Douglas (2021: 119–120) described how touring becomes the "new normal" and can lead to "disassociation" for crew members. She aligns these factors with the challenges crew members encounter when leaving the road. Douglas outlined how the insular nature of touring and the shared bond and understanding between crew members work to alienate them from "everyday people." The separation is further compounded by the nature of communication between crew and members of the non-touring public, which is usually in the form of requests and can cause greater isolation (ibid.). Touring "becomes your whole life" and a "lifestyle" that, for some, may also fulfil other areas, such as a sense of family life. After time, crew members adjust and take it for granted.

In this way, the people whom the crew member leaves behind adapt during their absence and must also reacclimate to their presence. As such, it is a mutual process in which both parties adjust. Joe (2017) stated that coming home from being a TM or a PM, roles that involve being in charge of a group of people, can create problems when coming home and trying to "be in charge of your household." He stated that his wife "hates it" when he returns from tour and that he is a "nightmare" for the first couple of days. He added that he "had to learn that I'm not in charge at home" when he returns from tour (2018).

> Doesn't usually go down well … You're coming into this nice settled domestic scene, who's been nicely managing without you, you know … and you come in and you want to be involved, you want to help and you want to do all these active things you've been dreaming about on the road, lovely long cycle rides with your happy family all smiling and you know and they tell you to piss off cause there's washing to do and there's badminton club … So that routine is quite difficult.
>
> *(2017)*

The kinds of activities described here suggest that the home is viewed as a holiday rather than the site of everyday life. This perspective was further evident in a statement that Tony (2018) made about his home being a "holiday place" and that his "life is not at home." Rather, home is idealised as a place to reconnect with family members. This suggests the kind of detachment crew members feel from home and reconfirms the manner in which they compartmentalise their lives. The need to engage in the everyday activities of home life comes as something of a shock to the crew member. Like Joe, Tony (2017) connected the challenge of adjustment to his role on tour, in his case, as close protection security in which he is continuously on call for an artist and builds his schedule around the musicians' lifecycle.

> … but you come in, you have been at somebody's beck and call for months, 24/7 for months and then you … home and then your wife will

> say the garden needs to be ... and you become quite resentful that the garden needs to be tended, which is a bizarre scenario in itself, it's your garden and you're not being asked something unreasonable, it becomes more about the psychological impact of not being able to, make your own decisions if you like at times.

In this case, the frustration comes from a continued pattern of being needed or expected to do something. It suggests the need for a break from the demands of the role and that certain aspects of home life can reproduce the expectations on tour. In these ways, roles may influence the nature of the challenges of coming off the road. This chapter will now close with a summary and a conclusion.

Conclusion

This chapter has analysed the complex mobile world of touring and has shown the ways in which it significantly shapes the working lives of road crew members. The activities and efforts involved in being on tour are constitutive, and integral to the realisation, of live music events. This means that concerts are the result of a long-hours culture, which is the product of the length of a show day, the travel between points A and B and the management of work relations during that time. Live events are therefore constituted by the complex working relationships between crew members. The latter are maintained by impression management and mutual respect, which crew members rely on for the acquisition of future work and protect their prospects by fitting in with the norms and expectations of touring. Concert events are further realised by both the challenging conditions of touring and road crew members' attempts to cope with them through shared workplace humour and an intense sense of camaraderie. However, these efforts are also complicated by the potential for bonds to be ephemeral and the implications of friendship on the road. As live music is a structure that distributes roles, the realisation of concert events is also marked by differences between them. The workplace culture of touring is permeated by masculinity and reproduces gender inequality and stereotypes. This can be observed through the expectations around conduct and the discouragement of weakness, and also at the intersection of mental health, in which women occupying particular roles become caretakers of others. After concerts have been realised, the process of touring continues to shape road crew members' lives as they encounter difficulties adjusting to the patterns of daily life off the road. The challenges of adaptation imply that touring functions as a stabilising feature that can foster a sense of normalcy. In this way, road crew members' lives are shaped by the experiences of being both on and off the road. The next chapter will focus on the role of tour managers and their working relationships with musicians.

Notes

1 Available at https://www.tourmgmt.org/templates.html
2 Tour books may not include the contact information for artists due to privacy concerns (Workman 2012: 146).
3 The costs associated with touring are outside the scope of this book. See Reynolds (2008), Waddell, Barnet and Berry (2007) and Nóvoa (2012). Documents in the David Russell Collection at the Rock & Roll Hall of Fame Library and Archives also provide detailed evidence of expenses, particularly for tours of top-level musicians.
4 Gorman (1978: 17) provides another example of how touring inhibits road crew members from having the "time to stop and look."
5 Episode 158 of the podcast *Roadie Free Radio* features an interview with the members of the Tour Health Research Initiative and information about the survey and its goals. It is available at: http://www.roadiefreeradio.com/podcast-1/2020/2/10/tour-health-research-initiative
6 https://bectu.org.uk/
7 https://www.helpmusicians.org.uk
8 https://themmf.net
9 http://www.musicsupport.org
10 https://www.lighthopelife.org/tour-support

References

Adey, Peter. 2017. *Mobility*. 2nd ed. London: Routledge.
Ames, Richard. 2019. *Live Music Production: Interviews with UK Pioneers*. New York: Routledge.
Banks, Mark. 2007. *The Politics of Cultural Work*. Basingstoke: Palgrave.
Barker, Emily. 2017. "Forgive the Rock Star Who Gets Your City's Name Wrong. They Probably Haven't Slept." *The Guardian* (16 February): https://www.theguardian.com/commentisfree/2017/feb/16/forgive-rock-star-city-wrong-guns-roses-axl-rose-melbourne-sydney.
Britton, Luke Morgan. 2015. "Insomnia, Anxiety, Break-Ups … Musicians on the Dark Side of Touring." *The Guardian* (25 June): https://www.theguardian.com/music/2015/jun/25/musicians-touring-psychological-dangers-willis-earl-beal-kate-nash.
Clawson, Mary Ann. 1999. "Masculinity and Skill Acquisition in the Adolescent Rock Band." *Popular Music* 18(1): 99–114.
Cody the Roadie. 2020. "The Bus Rules." *This Tour Life*: https://thistourlife.com/the-bus-rules.
Collinson, David L. 1988. "'Engineering Humour': Masculinity, Joking and Conflict in Shop-Floor Relations." *Organization Studies* 9(2): 181–199.
Cook, Malcolm. 2011. *Cook's Tours: Tales of a Tour Manager*. York: Music Mentor Books.
Cooper, Leonie. 2007. "Ladies Who Lug." *The Guardian* (13 April): https://www.theguardian.com/music/2007/apr/13/popandrock.gender.
Cresswell, Tim. 2006. *On the Move: Mobility in the Modern Western World*. New York: Routledge.
de Certeau, Michel. 1984. *The Practice of Everyday Life*. Translated by Steven Rendall. Berkeley: University of California Press.

Douglas, Tana. 2021. *Loud: A Life in Rock 'N' Roll by the World's First Female Roadie*. Sydney: ABC Books, HarperCollins Publishers.
Durkheim, Emile. 1983. *Pragmatism and Sociology*. Cambridge: Cambridge University Press.
Edgar, Robert, Kirsty Fairclough-Isaacs, Benjamin Halligan and Nicola Spelman, eds. 2015. *The Arena Concert: Music, Media and Mass Entertainment*. New York: Bloomsbury Academic.
Embleton, Tamsin, ed. 2023. *Touring and Mental Health: The Music Industry Manual*. London: Omnibus Press.
Evans, Paul and Jonathan Green. 2017. "Eyes Half Shut: A Report on Long Hours and Productivity in the UK Film and TV Industry." Broadcasting, Entertainment, Communications and Theatre Union (Bectu).
Feldman, Daniel C. 1984. "The Development and Enforcement of Group Norms." *Academy of Management Review* 9: 47–53.
Felski, Rita. 1999. "The Invention of Everyday Life." *New Formations* 39: 13–31.
Frankel, Jillian. 2017. "Watch Guns N' Roses Get Booed for Forgetting the City They're Playing In." *Billboard* (14 February): https://www.billboard.com/articles/columns/rock/7693236/guns-n-roses-booed-forgetting-city-melbourne-sydney.
Frith, Simon, Matt Brennan, Martin Cloonan and Emma Webster. 2019. *The History of Live Music in Britain, Volume II: 1968–1984*. London: Routledge.
George, Ryan. 2017. "The Mental Health of Tour Life. On and Off the Road." *This Tour Life* (17 January): http://thistourlife.com/mental-health-tour.
Goffman, Erving. 1959. *The Presentation of Self in Everyday Life*. New York: Anchor Books.
Gorman, Clem. 1978. *Backstage Rock: Behind the Scene with the Bands*. London: Pan Books.
Gross, Sally Anne and George Musgrave. 2020. *Can Music Make You Sick? Measuring the Price of Musical Ambition*. London: University of Westminster Press.
Hand drawn comics, "Road Rage," David Russell Collection, Library and Archives, Rock and Roll Hall of Fame and Museum.
Hawes, Kim. 2019. *Confessions of a Female Tour Manager*. Independently Published.
Hennion, Antoine. 1989. "An Intermediary between Production and Consumption: The Producer of Popular Music." *Science, Technology, & Human Values* 14(4): 400–424.
Holloway, J. Christopher. 1981 "The Guided Tour: A Sociological Approach." *Annals of Tourism Research* 8(3): 377–402.
Kearney, Mary Celeste. 2017. *Gender and Rock*. Oxford: Oxford University Press.
Kielich, Gabrielle. 2021. "Fulfilling the Hospitality Rider: Working Practices and Issues in a Tour's Supply Chain." In *Researching Live Music: Gigs, Tours, Concerts and Festivals*, eds. Chris Anderton and Sergio Pisfil, 115–126. London: Taylor & Francis/Routledge.
Leary, Mark R. and Robin M. Kowalski. 1990. "Impression Management: A Literature Review and Two-Component Model." *Psychological Bulletin* 107(1): 34–47.
Leed, Eric J. 1991. *The Mind of the Traveller: From Gilgamesh to Global Tourism*. New York: Basic Books.
Leonard, Marion. (2007) 2017. *Gender in the Music Industry: Rock, Discourse and Girl Power*. Aldershot: Ashgate Publishing Limited.

McRobbie, Angela. 2002a. "Clubs to Companies: Notes on the Decline of Political Culture in Speeded Up Creative Worlds." *Cultural Studies* 16(4): 516–531.
McRobbie, Angela. 2002b. "From Holloway to Hollywood: Happiness at Work in the New Cultural Economy." In *Cultural Economy*, eds. Paul du Gay and M. Pryke, 87–114. London: Sage.
Meintjes, Louise. 2003. *Sound of Africa!: Making Music Zulu in a South African Studio*. Durham, NC: Duke University Press.
Murphy, Claire. 2019. *Girl on the Road: How to Break Into Touring from a Female Perspective*. Independently Published.
Nóvoa, André. 2012. "Musicians on the Move: Mobilities and Identities of a Band on the Road." *Mobilities* 7(3): 349–368.
Practical Jokes, David Russell Collection, Library and Archives, Rock and Roll Hall of Fame and Museum.
Reiner, Rob, dir. 1984. *This Is Spinal Tap*. Spinal Tap Prod. Embassy Pictures.
Reynolds, Andy. 2008. *The Tour Book: How to Get Your Music on the Road*. Boston, MA: Cengage Learning.
Schuchat, Molly G. 1983. "Comforts of Group Tours." *Annals of Tourism Research* 10: 465–477.
Simpson, Dave. 2015. "How to Put on a Mega-Gig: The Roadie's Story." *The Guardian* (26 June): https://www.theguardian.com/music/musicblog/2015/jun/26/how-to-put-on-a-mega-gig-roadie-jon-bon-jovi.
Thomas, Jawsh. 2020. "Roadie Dictionary: A List of Touring Terms." *Backstage Culture*. https://www.backstageculture.com/roadie-dictionary-a-list-of-touring-terms.
Tolhurst, Lol. 2016. *Cured: The Tale of Two Imaginary Boys*. London: Quercus.
Vilanova, John and Kyle Cassidy. 2019. "'I'm Not the Drummer's Girlfriend': Merch Girls, Tour's Misogynist Mythos, and the Gendered Dynamics of Live Music's Backline Labor." *Journal of Popular Music Studies* 31(2): 85–106.
Waddell, Ray D., Rich Barnet and Jake Berry. 2007. *This Business of Concert Promotion and Touring: Practical Guide to Creating, Selling, Organizing, and Staging Concerts*. New York: Billboard Books.
Williams, Raymond. 1983. *Keywords: A Vocabulary of Culture and Society*. Revised ed. New York: Oxford University Press.
Workman, Mark. 2012. *One for the Road: How to Be a Tour Manager*. Road Crew Books.

Interviews

Adrian. Guitar Technician. In-Person Interview, 24 May 2018.
Amy. Tour Manager. Skype Interview, 9 July 2018.
Andy. Tour Manager and Sound Engineer. In-Person Interview, 18 April 2017.
Tasha. Crew Member. Phone Interview, 14 August 2018.
Craig. Monitor Engineer. In-Person Interview, 4 May 2017.
Duncan. Guitar Technician. Phone Interview, 10 May 2018.
Joe. Tour and Production Manager. In-Person Interview, 2 May 2017, 14 May 2018. Skype Interview, 28 August 2019.
Tony. Close-Protection Security, Head of Security. Skype Interviews, 4 August 2017, 16 March 2018.

5

LOOKING AFTER MUSICIANS

As the name suggests, the tour manager is responsible for managing the tour. If live music is the purpose and daily objective of a tour, musicians are its central component. In this way, the most significant aspect of the TM's job is to look after musicians. The TM is the "person who goes on tour to ... look after the administration of the band day-to-day" (Andy 2017). A TM at Venue A summarised his job as being "primarily" based on managing musicians while the author of a guidebook on tour management stressed that "first and foremost, a [tour] manager gets paid to take care of the band" (Workman 2012: 312). To "look after" or "take care of" artists is to effectively attend to the necessary tasks that relate to their working lives on tour that enable the realisation of live music events. The proximity in which TMs work with musicians symbolises this vital aspect of their roles. They work alongside musicians throughout the duration of the tour, travel in the same transit vehicle with artists and accompany them to venues and on off-site promotional tasks. On show days, the efforts of TMs have been realised the moment the house lights go down, the crowd cheers and the artist walks onstage.

This chapter examines how TMs look after musicians and the attributes of their working relationship. It highlights the important role of care in the working relationship between TMs and musicians and how this component of the former's job is intertwined with their occupational identity. The chapter demonstrates that the tour manager's efforts in relation to musicians are a significant factor in the realisation of concerts and the daily practices of live music. It also shows that the parameters of the TM's role are adjustable and, at times, ambiguous and provides insight into the type and nature of activities that mark their working relationship with musicians.

DOI: 10.4324/9781003303046-6

The chapter begins with a discussion on the various factors involved in defining the TM's role and outlines key characteristics and their significance. It then moves to analyse the ways that TMs look after musicians on tour through an exploration of essential daily activities. The section that follows examines the use of the term "babysitter" that is commonly applied to TMs to gain further insights about the work they do in relation to musicians.

Defining the Role

The role of the TM is versatile. It is neither standardised nor presented in a formal job description and the exact responsibilities vary with and are shaped according to context. The role is defined, first, on the basis of the size and requirements of a tour, and therefore its division of labour, as stated in Chapter 1. Second, it is determined by the specific needs and preferences of the artist for whom a TM works. In this way, their roles are *artist specific* as they are defined by and change according to the particular musicians for whom they work on a given tour and to which TMs correspondingly adjust their practices. Tour management is essentially a custom-designed service based on individual artists and the features of their tours. Third, the role can consist of a wide range of tasks and responsibilities that vary in kind and shape the working lives of TMs (see Hart 2011: 1). In this way, TMs negotiate and establish boundaries to define the limits of their roles prior to accepting a position and the start of the tour.

TMs view flexibility as a necessary attribute of their roles and artists as a determining factor in defining them. Adaptability is an asset as it allows for differences between musicians and creates the possibility of tailoring the role. TM Joe (2019) does not "think my role should be heavily defined as in 'a TM does this.' Yeah, I don't think a TM should always do this, I think you should define your own role with whatever artists." Former TM Mark Workman (2012: 315) explained that every artist is "different from the last. Each one of them is a different personality with different needs and expectations." TM Amy (2018) likened the role to

> project manager would be like the best description … you're kind of overseeing a lot of people, you're making sure all the components work … it's kind of a little bit of psychologist and stuff thrown in … because that is part of project management to handle different personalities.

Equating tour management with project management substantiates that the former changes on a case-by-case basis. These quotes confirm the artist-specific nature of the role and underscore its adjustability. Research on support personnel in the film industry helps to further illustrate the malleable nature of their roles. Makeup artists, for example, as the first people who work with actors every day "must also manage each performer's professional concerns and personal sensitivities, sometimes playing the role of confidante and therapist while also crafting a countenance that will be projected across millions of screens" (Curtin and Sanson 2017: 3). Such "excessive labour" is the "hidden, voluntary, unrecognized, and often unwaged aspects" (6) – and "persistent pressure for 'more'" – that can characterise work in the cultural industries. Workers in the film industry see such responsibilities as an unavoidable part of their jobs and "grudgingly accept" them (7).

The notion of excessive labour highlights the possibility for such types of activities to infiltrate the working lives of cultural workers and calls attention to the array of tasks that are unknown or taken for granted. However, the idea of excessive labour is useful for understanding tour management not because it is the same but due to the fact that it is different and does not easily map onto the ambiguous job description of a TM. While in the film industry such activities are deemed in excess of usual job duties, for TMs, the types of activities and responsibilities, and the degree to which they take on particular tasks is defined and determined. In other words, what may be considered an act of excessive labour to a makeup artist could be understood as a normal part of the TM's role and would not be considered in excess of their job expectations. As such, TMs do not view nor accept those tasks as unavoidable. The potential for excessive labour does exist, and the nature of it falls into similar categories. The difference is that the definition is fluid and efforts are made to prevent excessive labour through advance negotiations, as this chapter will discuss. The nature of their roles means that what is in and outside of acceptability changes with context. This is related to the TM's own comfort level and tolerance for specific responsibilities and personalities, and to the boundaries they set and learn to determine.

Defining the role is significant because expectations influence the TM's working practices. Artists may want or need more involvement from the TM. Others are more "self-sufficient" in research respondents' terms. Different types of demands may be placed on the TM in terms of the nature of activities and interpersonal conduct. In terms of working with musicians, the aspect of the job that varies and requires defining is typically related to personal considerations. TMs expressed a range of views as to what is acceptable and expected of them.

> ...it'll often be on the more personal side of things, like ... you need to be a TM but on your days off you're expected to go shopping or you'll be expected to look after the artist in this way or that way ... Or ... this artist is explosive and difficult, can you cope with that, can you be a punching bag without just losing your shit and shouting at them? You will be expected to be a brick wall that gets shouted at occasionally.
>
> *(Joe 2018)*

Amy (2018) took a clear position on the personal elements she considers to be the TM's job, which implies that she believes that she has a degree of responsibility for musicians.

> Aside from business aspects of the gig, it's our job to keep an eye on the people on the tour and know what's going on, assist with resolving any conflict, get people out of sticky situations if they get in one, try and make sure they get some sleep, call doctors if they're sick.

Such activities require a much more personal involvement on the part of TMs in maintaining the everyday lives of musicians. The notion that support personnel must do whatever it takes to make a show happen (see also Kielich 2021) extends into the interpersonal context in such cases. However, TMs may change their perspectives during their careers, and perceive musicians as being responsible and accountable for their own lives. The extent to which TMs believe they need to monitor musicians' behaviour may diminish with experience.

> ... as a TM of a young, successful band who are having a great time, I thought it was my job to crack the whip and make sure that ...you know, go and drag them out of clubs when they had to get on a bus, that kind of thing, cause it was my responsibility to make the show happen. What I never took on board for a long time was that if they want to stay out partying until 5 and then the bus is four hours late the next day that's their fault, not mine. If they can't get up and get on a plane and you have to buy new plane tickets for 12 people because your transport hasn't been able to leave until they got on it, that's their fault, not mine.
>
> *(Joe 2019)*

Such activities that were previously seen as an essential part of the job later come to be recognised as outside its expectations. The views of artist management may also influence the TM's tasks in the personal aspects of working with an artist. Some may state their opinion while others will not; in the case of the former, "some would say 'no that's your responsibility get them on that plane' whatever. So there's an elastic dynamic between management, tour management and the artist that you have to find the balance is right" (Joe 2019). The ability for artist management to overrule the TM highlights the ambiguous relationship the latter have with power and authority, which will be discussed in more detail. Regardless of the nature of their artist-specific roles and expectations, it is ultimately the TM's decision as to whether they are willing to adapt through their choice to accept the position or not. "I think if you get warned about it in advance, you've signed up for it ... Take the job or don't take the job, it's up to you" (Joe 2018).

The parameters of their roles are initially discussed and determined during the hiring process in consultation with artist management. It is necessary to briefly explain the latter role in order to understand and differentiate it from the TM. The artist manager oversees all aspects of musicians' careers (Reynolds 2012: 154). They represent artists in all business affairs and advise them on logistical and financial decisions (Reynolds 2008: 12). Musicians hire them as a competitive strategy and in order to gain access to music markets (Jones 2012: 96). Artist managers are not fully involved in the daily activities of a concert and are typically not part of the touring party but may attend specific shows (Reynolds 2008: 11). Returning to the hiring process, the TM's goal is to become aware of an artist's particularities, understand the types of activities that will be required and establish the acceptable tasks and limits of the role. Amy (2018) stated that TMs must "draw your lines, draw them early and don't cross them." This is particularly important because "especially [with] touring, whatever you offer to do" will become part of the TM's job (ibid.). Workman (2012: 313), in his guidebook about tour management, similarly expressed that TM's "should be honest up front" about what they are and are not willing to do before being hired. They should also avoid making assumptions about the types of duties an artist expects (ibid.). Artists may be accustomed to the practices of a previous TM, but a new person in that role may have a different approach. Research respondents explained that artist management will usually offer some insights and guidance, but that it is fundamentally dependent on TMs' initiative to ask questions.

A serious and extreme example demonstrates how the artist-specific nature of the role, the need to set boundaries and the degree of a TM's comfort level come together and operate in practice to define the role. TMs may be offered positions that would involve working for artists with substance abuse problems. The topic of alcohol and drugs as a component of musicians' lifestyles has been minimally examined (Raeburn 1987, Groce 1991, Forsyth, Lennox and Emslie 2016, Bennett [1980] 2017). It has not, however, been considered as a mediating factor between musicians and the personnel who work around them. Working for artists with substance abuse problems can create a stressful and exhausting work experience for TMs. Early in his career, Joe (2017) worked for an artist with long-standing and well-known addictions. He recounted the experience of ultimately choosing to leave the tour due to the difficulties it created.

> ...till I couldn't take it anymore. His substance abuse or addiction is not a secret. And just exhausting dealing with it ... Yeah. It's just tiring, you know. I very quickly said to him, and he's been touring for years ... 'if you want to do that, I'm not getting it for you.' Long conversations with the management who had taken him to Harley Street* and they'd looked at getting him clean for years, and he was just having none of it. I don't think he was

> ready for it at the time. His alcoholism alone is pretty extensive, so ... one show in three would be spectacular ... And one show in three would be a car crash cause he was just in the wrong cycle of up and down ... storming off the tour a couple times and I'd talk him back, I'm quite good at that. It just became tiring. Didn't want to do it anymore.

Workman (2012: 312) wrote that dealing with an artist's substance abuse can be a "scary situation ... especially when you're in charge of taking care of them." Part of this is related to how TMs are not "drug and alcohol counselors[s]" and are likewise not equipped with the methods or trained to address the needs of people confronting such difficulties (313). TMs also potentially put their legal status at risk if they are involved in the process of acquisition of illegal substances.

The close proximity in which TMs work with musicians can also raise questions about responsibility and the degree and nature of their involvement in aspects related to substance abuse in artists' lives. High-profile cases such as Amy Winehouse have generated debates about the extent to which people around an artist, including those in management positions, can seem enabling, neglectful and ultimately exploitative by failing to help musicians get the care they need or recognise the extent of their problems (see Gross and Musgrave 2020: 2, 130). Scholars have noticed consistent viewpoints held by workers in the music industries that

> imply that the health and wellbeing of musicians was not just their own personal responsibility but should ... be shared by the formal organisations and structures of the music industries ... and that this should be embedded within working practices.
>
> *(ibid.: 129)*

It has even been debated whether a strict, all-encompassing legal liability that covers all personnel working directly with artists should be put in place in the form of a "duty of care" (128–134).[1] The risk inherent in the music industries, the nature of long-term working relationships and the unpredictability of musical work make the development of best practices a more likely solution, however (134).

The working relationship between TMs and musicians on the road puts them in direct contact on a daily basis. Joe (2018) spoke to having reflected on such concerns in relation to his own role. He described "putting myself in [Winehouse's TM's] situation," contemplating how he would feel and asking himself a variety of questions regarding responsibility, awareness and intervention. The distinction between "trust" and "care" usefully characterises the TM's position in such cases. A "climate of trust assures that persons will have the right trustworthy intentions" but does not make certain that they will do what is needed nor does it provide that "persons are proficient at meeting the needs of the vulnerable" (Held 2006: 57).

Research respondents agreed that TMs are ultimately not responsible for protecting or enabling artists in such circumstances. Amy (2018) stated that "if you're knowing that [a musician's] gotta take something to get onstage, I don't think it's a TM's responsibility." Her approach to dealing with such situations is to try not to engage in the first place. She "just won't take the job. Like if I know somebody is ... well known as being a drug addict or just whatever, I would just be like no." Choosing not to work with such artists prevents any risk to both parties. Joe (2018) says the "best thing I think to do in those situations is openly discuss it with management, who will be aware of the situation. I think it's much more of management's side of things whether they intervene." Workman (2012: 313) also does not "believe it's the tour manager's job to play wet nurse to a musician who's abusing drugs or alcohol" but notes that it "usually ends up in the tour manager's lap whether he likes it or not." The artist management may be aware and able to offer guidance but "when it comes down to it, they're in an office in London and you're in South America" (Joe 2018) meaning that on tour it is the TM who is ultimately the person working with an artist on a daily basis.

Rather than TMs directly intervening, should artists have substance abuse problems any course of action "has to be artist-led, they have to make the decision that they want to change their behaviour if it's not working for them" (Joe 2018). As the support personnel hired specifically for the purposes of a tour, TMs may subtly call attention to the issue if it is having an effect on the concerts themselves.

> So you talk to them and see what they want and then try and support it. It's difficult from your point of view to point out that the shows are going a bit raggedy, might be worth taking [artist] to the side and going 'how you feeling, that wasn't the best was it?' Try and get them to make that decision themselves. But, yeah, you just have to ... get them from show to show with as little fuss as possible. It doesn't involve taking their booze away if they're asking for it.
>
> *(Joe 2018)*

TMs' responsibilities ultimately are based on looking after musicians in order to ensure that live music events happen. In this way, the specific habits of an individual musician only require attention if they negatively impact the quality of the shows and the continuity of a tour, of which TMs are in charge. Trying to intervene or mitigate more generally could also risk or compromise the ability for musicians to do their jobs successfully, which could create further problems for the tour. The issue of substance abuse further illuminates the difficulty of and significance in defining the TMs role and its boundaries. It also further clarifies that the TM's role is artist-specific and determined by their own willingness to adjust, and by how they choose to or refrain from working with particular musicians.

When TMs take a position, they must get to know and understand the particular musicians for whom they work. This practice is an extension of the artist-specific nature of their roles as it allows them to more accurately tailor them to meet the artist's needs. As they come to understand the musicians for whom they work and can base their practices around them, knowing an artist well enables greater proficiency and efficiency in their working lives. TMs try to "understand their psyche, what do they need from me as a …tour manager" (Joe 2017). Doing so is also an investment that creates greater potential for a sustained working relationship. To "last with any band, you must learn what makes them happy and content and, even more so, what does not" (Workman 2012: 315). TM Kim Hawes (2019: 345) described how she learned how "they liked to work, how they performed, what went into a show, how their tours were organised and who was responsible for what" during her lengthy working relationship with Motörhead. She came to know them "first-hand" and attributed this to the trust and capability she was accorded (345–346). She noted that because she had toured with various musicians, she was aware of what Motörhead "did in common with other bands and what they liked to do their own way" (345).

Joe (2019) described the process of making travel decisions and arrangements in coordination with the artist he works for.

> They're often very compartmentalised about what they're thinking about … there'll only be very specific times when you can talk to her about this stuff. She'll go 'no I can't think about it just now, can't think about it just now.' And she's not necessarily doing anything else, but … she'll go 'no, my head's not there.'

Joe learned to discern when the artist is ready to have such conversations and approaches her to discuss it then. This demonstrates that TMs tailor their practices around artists and that they come to do so by getting to know them.

> And then I can recognise when she's okay and I'll go '… can I talk to you about this' and then she'll go 'right, go.' And … it'll be something about six months away, 'right, we've got two days off there, would you rather go to New York or Boston' or whatever it is and she'll try and put herself in that place and think about it and give me an answer to that.

While TMs are responsive to artists, they also work to foresee their needs and prevent problems from arising. The better a TM knows and understands an artist the more it is possible to anticipate their needs, manage their expectations and troubleshoot problems, and to do so before or even without musicians being aware or needing to know. A TM is "not only a problem solver but a problem detective" (Workman 2012: 315) and has a responsibility to "try to foresee problems arising before they actually occur" (Cook 2011: 78, also Hawes 2019: 346). Working to get to know artists "better than [they] know [themselves]" means they can "detect when a problem is brewing ... long before it manifests" (Workman 2012: 315). Rather than waiting for the artist to tell TMs what they need, TMs should "[f]igure out what [the artist's] personal needs are" by themselves and ensure they are dealt with (ibid.). Knowing the artist well facilitates greater initiative on the TM's part by allowing them to act without needing to consult the artist first and therefore contributes to the overall smooth running of the tour. Gerry Stickells, the longtime TM of Queen, described, prior to his death, that understanding and acclimating to an artist's disposition means "you can usually stay a step ahead" (Sandomir 2019). TMs do so in accordance with the artist's needs and preferences, but actions and decisions are also based on what the TM *perceives* the artist wants, or on the TM's own inclinations for dealing with particular situations. Their working relationship can involve the negotiation of a power dynamic that is based on musicians' preferences and the TM's assessment of a situation.

A successful working relationship depends on creating common ground and mutual understanding. The ability to build trust between the TM and the artist is an essential component. Joe (2017) recalled an example in which a self-managed duo he worked for provided lists of all tasks they wanted him to do and would check it over with him. He explained that they would ask "'have you done A, B, C, D, now you're on E.' 'No I haven't done E,' 'why not?'" Their actions implied that they questioned the TMs' authority and competence, and their lack of trust made the working relationship difficult and unbalanced. In another case, Joe worked with an artist who he stated "fully" understood the relationship between the TM and musician.

> ...he really appreciated what I was doing and if I said to him, 'no we need to do this today and not that' he'd go 'yeah whatever, you've made the choice I trust you.' Sometimes we'd chat about it afterwards but he wouldn't question what I was doing. And I think we fell into that fairly quickly, I think he found that he liked what I did so it just clicked.

The working relationship between an artist and TM works best when the former believes the latter to be competent and trusts the decisions that are made. TMs try to achieve balance and understanding while maintaining authority and autonomy. Trust between them is an essential factor that shapes their working lives and can make the TM's job more or less difficult.

Such examples also highlight that TMs have an ambiguous relationship with power and authority. On the one hand, this can be understood in relation to their position as managers. On the other, the role of the TM can be equated with "middle management" in an organisation. Generally speaking, middle managers are "located below top managers and above first-level supervision in the hierarchy" (Wooldridge, Schmid and Floyd 2008: 1192). In the context of a tour, the TM works for the artist, and though the latter are not managers, their centrality to the live music event and role in shaping the TM's job put them in a position of influence. The TM reports back to artist management and oversees the rest of the touring party, though, as stated, roles can be further divided depending on the size of the tour, as when production-related tasks are delegated to the PM. TM Andy (2017) described the role as "I'd say you're not a manager in terms of a position of authority like in a more conventional workplace, you are managing the activities and the timeline … and the money." As example, the TM has responsibility for expenses associated with the approved budget, but if an emergency occurs, the TM should "definitely seek the [artist] manager's approval before spending any more money" (Reynolds 2008: 16). Even though the TM does make important financial decisions, the artist manager oversees expenses incurred (ibid.). Joe (2019) described a relationship with management that contrasts slightly with that outlined in the statement by Reynolds. The latter's statement was primarily in relation to new artists while Joe works for established musicians, which suggests the former needs to be more cautious and that TMs are granted more autonomy with experience. As middle managers, TMs can and do have a degree of autonomy, and are trusted in decision-making, yet artist management may want to simply verify or maintain awareness of the rationale.

> … I can spend money without having to have everything approved, they trust me just to get on with it and make it work, and 'oh suddenly we're 50,000 over budget,' they might ask me why and I'll say 'this is the reason' and they'll go 'fine.' … my autonomy on the road isn't questioned.

More important than hierarchy to middle management is their combined access to top management and knowledge of operations (Nonaka 1994). This combination means they "function as mediators between the organization's strategy and day-to-day activities" (ibid.). TMs manage expectations and handle the challenges and uncertainties of everyday life while working to realise the goals of musicians and the requests of artist management. Clem Gorman (1978: 39) gave a relevant example of a TM who described that he gets

> involved in management-type decisions sometimes, in fact I'm the middleman between management and road crew, and sometimes between

> management and band, because I have to have a say in what venues we can and cannot do, whether they're too small, acoustically unsuitable, or whatever.

Similarly, middle managers "work as a bridge between the visionary ideals of the top and the often chaotic reality on the frontline of business ... They even remake reality according to the company's vision" (Nonaka 1994: 32).

TMs ultimately move between positions of authority and subordination. Like middle managers, they have some ability to influence decisions and, as stated, negotiate their job content (Kankkunen 2014: 341). As a type of middle manager, their working lives are marked by "role conflict" given the relative incompatibility of the norms and expectations associated with being a leader with those of being a subordinate (Anicich and Hirsh 2017a, b). Such conflict or contradiction in the notion of management was also observed by Raymond Williams (1983) and applied to artist management by Mike Jones (2012: 89). It is also useful for understanding TMs. Williams (1983: 189) found that the first usage of the term was from the Latin *manus* ("hand") and was adopted in English from "maneggiare" a medieval Italian word that refers to handling horses or driving horse-drawn vehicles. This can be linked to such terms as "taking the reins of power" and "being in the driving seat" (Jones 2012: 89). The second usage is also from Latin and comes from *mansionem* or "a large dwelling or early form of hotel" and was adopted in English from the medieval French terms *ménage* or *ménager*, referring to "household" (Williams 1983: 190). This meaning of management aligns it with "housekeeping" or a more "hands-off" form concerned with attention to detail that differs from the "hands on" usage based on the idea of leading (Jones 2012: 89). The usages have merged, but each are still evident in practice (Williams 1983, Jones 2012). Jones (2012: 89) notes that artist management is "not a clearcut undertaking" as the role can involve the expectation to "be a leader while checking constantly that there are always guitar strings, valid passports; and that there is always money in the bank." Such contradiction in a context that is "light on rule-bound behaviour" can be "an unrecognised source of irritation and tension" (ibid.). In other words, these disparate sets of responsibilities put the TM in a position that moves between opposite ends of the hierarchical spectrum and can create difficulties in defining their occupational identities.

Middle managers have a "complicated relationship" with power because it is "activated and experienced in the context of interpersonal relationships" (Anicich and Hirsh 2017a, b). The relationship between the TM and musicians is marked by the potential for conflict between them based on the question of authority. In turn, this creates "role strain" on the former that can affect role performance. The TM is essentially hired to perform a specific service, which lends to the notion that "the customer is always right." As outlined in the Introduction, musicians typically have a higher social status than TMs. However, like that of a tour guide in the tourism industry, the "competent performance" of the TM's role relies on "temporary authority over his clients for the length of the tour, in a manner similar to that enjoyed by most professionals, such as physicians or lawyers" (Cohen 1985: 23). In the case of tour guides, the potential for conflict is based on an absence of clear professionalisation. For TMs, this would unlikely be a cause given the accepted informal practices in the music industries. Rather, it is based on the implicit effects of musicians' status and the payment arrangement between them. TMs are paid from the tour's budget, and their fees are understood as a reduction of the artist's earnings. Musicians hire and pay TMs directly and, in this sense, are relative authority figures as their "boss." Like artist managers, TMs are essentially "service provider[s] to musicians" and confront the "dilemma that a servant is hired to effectively be a leader" (Jones 2012: 90). This arrangement raises questions about authority and TMs and musicians may engage in power struggles "with no clear guidelines for their resolution" (Cohen 1985: 23). The financial arrangement and the working dynamic between TMs and musicians reveal that the relationship is marked by an inherent tension. Joe (2017) recounted an incident that illustrates this tension. He lost his job with an artist over a disagreement about a birthday party being held on the tour bus.

> I lost my job ... because I refused to let the guitar player have a birthday party with a load of strangers on the tour bus. I went 'no they're not coming on,' put my foot down, just went 'no.' I have a crew upstairs who need to go to bed for starters, they all start work at 8 in the morning and you want these guys on who we don't know on the bus all night, travelling to the next city. And then I'm gonna have to kick them all off in the morning and they're gonna ask me where the train station is.

This shows a clear disconnect between the responsibility of the TM, the interests of the tour and the desires of the artist. He was working in what he perceived as the best interest of the band and tour, but this was at odds with the artist's interpretation of the situation. As the tour bus functions as a temporary home, security concerns and the wellbeing of the rest of the touring party are essential considerations. Allowing strangers on the bus is a practice that is generally discouraged by TMs. However, if the touring party decides that visitors can be brought onto the bus, they should establish rules regarding why and to what extent guests are permitted (Atkins 2007: 257).

> I didn't believe that was the best thing for the tour, I think that would have been really disruptive, could easily potentially even resulted in a broken bus, a very pissed off bus driver, lost or stolen money or lost or stolen

> laptops or crew that could quit cause they didn't get a wink's sleep and they have a job to do in the morning.
>
> *(Joe 2018)*

He suggested a compromise with the artist instead, offering to arrange a hotel for the party and a train to the next city the following day, but the artist did not comply and "burst into tears." The artist resisted the TM despite the latter being hired to provide a service that is intended to work in the interest of the tour and the artist. TM Iain Williamson affirmed that it is "in the band's interest for the TM to be in charge" (Simpson 2010). A lack of clear definition of authority can present problems, and TMs ultimately use their best judgement without having a clear sense of the boundaries.

> They're the boss, it's their job, it's their gig, they're paying you, but ... I refuse to say yes to everything. I won't say yes to dangerous things, that's the hardest line and that's the nub of being good or bad at it, is finding that line. It moves, the line moves and you've got to watch it all the time and decide what you're comfortable with and what you aren't.
>
> *(Joe 2017)*

This also suggests that the grounds for firing a TM are unclear and subject to an artist's discretion. TMs are accountable for both the tour and the artist, and such an occurrence demonstrates that they can conflict. The fact that this role is defined in a relationship with the artist at once enables TMs to do their jobs while it can also present risk. Though they attempt to define their roles in advance, the expectations can be complicated by incompatible interests and interpersonal disagreements. Foregrounding the overall interest and safety of the tour and the touring party is the TM's responsibility; being dismissed for doing so suggests that the boundaries of the role are open to interpretation. TMs make their best judgements but the tension between providing a service and being a leader means they can be overruled by those in a more clearly established and consistent position of power. The next section will provide an overview of the significance of the TM's role to live music events.

The Significance of the Tour Manager's Role

The TM's role on a tour can be divided into three stages. The first is to organise the tour, which happens prior to the tour actually starting. TMs generally become involved once the tour dates have been booked, which is a process handled between the artist management and the booking agent. The booking agent is in charge of finding remunerated performance work for musicians, negotiating with promoters and booking concerts (Reynolds 2012: 156, 2008: 17, 19).

They are not amongst the personnel who comprise the touring party. This position works in liaison with the promoter and artist management to plan the specific dates. The next stage is to advance the tour, which is a dynamic process, as detailed in Chapters 1 and 3, that begins prior to the tour's commencement but continues on a daily basis as it progresses. Third are the everyday activities that take place on tour and the closely related working relationship with musicians during that time, which are the central focus here. TMs both formulate and implement aspects of the tour, which further reflects their role as middle management.

The previous section established that TMs' roles are artist specific insofar as they are shaped and defined in relation to specific musicians. Looking after an artist day to day means that TMs' jobs are also *artist centric*. The attendant set of tasks and responsibilities fall into four categories. TMs work to *provide* a conducive and comfortable environment for musicians on tour that enables focus on and preparedness for live performance. This includes the coordination and communication of all necessary information to facilitate the realisation of live music events. TMs *protect* musicians from conditions and circumstances that have the potential to negatively affect them. They *support* musicians in their endeavours, live music career goals, personal preferences and to meet their daily needs on tour. As some of their activities have effects that "actually sustain persons" and involve work that "responds to their vulnerabilities" (Held 2006: 57), TMs *care* for musicians to some extent. Joe (2018) summarised the goal of the TM's responsibilities as follows:

> Ideally you are there to facilitate their shows as smoothly as possible with as little friction, as little difficulty for the artist as possible. You need to get them and the crew and the equipment from place to place with the least fuss in the most efficient manner. And produce the shows that they want with the least fuss … to be as invisible as possible while negotiating and solving problems and generally making their life easier, easy as possible, take as much off their shoulders so that they only have to think about the show and doing a presentation.

TM Malcolm Cook (2011: 15) similarly characterised the role as being to "ensure the smooth running of the show … and generally ensure that the tour was as trouble-free as possible." TMs are hired to carry out duties that "free executants from normal household chores" (Becker 1982: 4). Making art – or playing live and touring – is time and energy intensive and "has to be diverted from other activities" (3). TMs alleviate an expenditure of energy and enable its direction towards live performance while also ensuring that musicians do not go without certain services and that necessary tasks are being dealt with. Brian Hracs (2015) has shown that entrepreneurial musicians encounter difficulties organising and designating time and energy to the core creative aspects of their working lives due to the number of tasks they must attend to. Applied to the live setting, musicians would risk compromising performance without the assistance of TMs. TMs create, and protect, the space for musicians to ensure their focus on and readiness for performance, and work to maintain consistency for the artist.

Tour management is "a compassionate understanding of the stress of being on the road" that can help, or hinder, an artist (Atkins 2007: 253). Joe (2017, 2019) characterised the "principal role" and a motivating factor of his job as being "the satisfaction about taking stress away from artists so that [the] performance happens." TM Paolo Francesco identified that the primary part of his role is "making sure your artist is comfortable while they are doing their job" (Raven 2015). In the same way that a porter bringing coffee to writer Anthony Trollope each morning was seen by the author as integral to completing his daily literary work (Becker 1982: 1), the variety of tasks TMs carry out are essential to musicians' working practices. As discussed in Chapter 1, in the early stages of musicians' careers, paying a percentage of earnings to road crew members is deemed worthwhile because it reduces the physical work of touring (Bennett [1980] 2017: 75–76). This pattern continues and evolves at later and more successful stages with a larger division of labour and the efforts of the TM. Musicians hire artist managers to "broker relations with music companies" and, in doing so, they transfer "much operational responsibility to them" (Jones 2012: 78). The same applies to TMs in the context of touring and in daily dealings with live music industry personnel. Musicians rely on TMs to attend to these essential tasks (Small 1999: 9–10), as does the live music industry. If the recording industry depends on the distance between creative managers and musicians to facilitate the latter's autonomy and innovation during the creative process (Hesmondhalgh 2013: 32, Stahl 2013: 233), the concert industry depends on the proximity of TMs to assist musicians for the success of live performance.

The centrality of touring to musicians' careers, as both promotion and remuneration, means live music is a pressurised workplace. Playing live is an important space in which musicians market recordings, test and develop new music and work to establish and maintain a fanbase (Black, Fox and Kochanowski 2007: 154–156, Shuker 2008: 55, 57). The lucrative economic value of concerts, ticket prices, competition for audiences and the site of live performance as a source of evaluation and authenticity are further contributing factors (Auslander 2008, Frith et al. 2013, McKinna 2014). Live performance itself is a demanding endeavour that must be repeated in different locations continuously throughout a tour (McKinna 2014). For musicians, it is a challenge to "maintain freshness at each performance" as each event can blend into the next (Shuker 2008: 57). Physical stress is also placed on musicians in the live setting, as was observed and described by TMs. During participant observation, a TM talking with a promoter cited the exhaustion that musicians experience after a concert as a reason that it is important for him to understand the artists he works for. Joe (2019) observed the intensity of live performance for various types of musicians.

> Physically it can be extremely demanding as well. You know, a two-hour show five or six times a week is can really take its toll. Especially for drummers and things like that that are just physically giving their all … I had [artist] at [venue] recently and … it got more and more intense and more and more intense as it went on and you know, the last few songs … she's just giving it her all and she had to be helped down the stairs afterwards, she didn't come across as that onstage but she left it on the field if you like and then, just coming off she's like 'just get me to a seat' and straight to the car and just like shattered, you know.

Simon Frith et al. (2019: 168–169) have shown that hard work is a common theme in rock memoirs and much of this discussion is in relation to touring and live performance. They highlight that the challenge of touring is not only the repetition of live performance but also the daily promotional efforts such as press interviews and photo shoots. In turn, the attention and anticipation generated by the press adds pressure on musicians to meet expectations in the live setting (ibid.). Furthermore, the demands of live music are distinct from those placed on other workers in the entertainment industries. Film actors may travel to work "on location" and work long and irregular hours, but they have the safety of retakes and edits. Theatre performance involves a gruelling schedule of eight shows a week, but actors have understudies should they fall ill or be absent due to other circumstances. Professional athletes also have replacements should they be injured or need to sit a game out. The same conventions are not part of the practice of touring. Musicians, in general, do not have backups. An inability to perform means cancellations or the complications of rescheduling shows. In these ways, looking after musicians and their needs is in the interest of the overall success of a tour. The next section will discuss the daily activities TMs engage in while looking after musicians.

Daily Activities on Tour

A TM's daily activities are very similar to those of a tour guide in the tourism industry. The context in which they work and the purpose of their tasks contrast, but they share essential responsibilities that shape their working practices. To frame them, this book adapts Erik Cohen's (1985) four major components of the tour guide's role to our examination of the TM. The first, the instrumental component, relates to the TM's purpose in ensuring the "smooth accomplishment of the tour" and in "leading the way" (11). After the advanced determination of the itinerary, the TM handles the "spatio-temporal direction" of the tour, which involves properly following its direction and adapting when necessary (ibid.). The duties that involve handling the tour's logistics means they are hired to guide the touring party for the overall purpose of realising live music events. Touring is a repetitive process that is structured and organised by an itinerary and operates most efficiently when carefully planned in advance. However, the process of continuously moving from city to city means TMs inevitably confront unpredictable issues on a regular basis, and an important part of their working lives involves troubleshooting them. TMs' handling of "seemingly unexpected events" and doing so with a particular urgency and minimal disruption to the continuity and atmosphere of the working day (Tuchman 1973: 111, 117, 119).

Along with this, the TM is responsible for the "safe and efficient conduct" of musicians (12). On the one hand, this means ensuring, to the degree possible, their security, safety and comfort. On the other, the TM should "exercise control" by preventing them from breaking away from the tour or group, collecting stragglers and monitoring the pace of movement (ibid.). TMs have to be concerned about musicians' whereabouts because live music events depend on their presence. In the tourism industry, this practice is sometimes called the "shepherding and marshalling" (Holloway 1981: 380) function of the guide. TMs engage in similar practices when necessary. Joe (2018) described having to "chase" musicians when they became distracted by local amenities near a venue. Former Coldplay TM Glen Rowe expressed a similar experience on his Twitter account when he described being in New York City with a different artist and having "no idea where any of the band members are." The practice is "very much like herding cats. It's like every time you turn around and one of them is wandering off doing something" (Joe 2018). Former backline technician Ken Barr (2009: 51) similarly characterised TMs as "like the farmer chasing all his chickens and trying to get them together." Such circumstances suggest a sense of distraction on the part of artists and create additional tasks and greater obligation in the daily working practices of TMs. In this context, their function as leaders or guides is to some degree compromised and their role becomes a struggle to control or hold a situation together. The role of the runner, a member of the local crew who assists the touring party with off-site needs, transport and access to and around the local area (Kielich 2021: 120), functions as an extension of this responsibility when accompanying musicians in the local area. The runner is knowledgeable about the local area and is the touring party's connection to it, which means it is both logical and convenient that they should accompany musicians. However, it is also another means by which the latter are "marshalled," and knowledge of artists' whereabouts is maintained. It is a way of knowing where they are and ensuring that they will return. At Venue B, a runner accompanied the lead singer and keyboard player to visit the city, and it was announced in the green room when the musicians returned.

A more extreme example involved an early tour during which Joe (2017) described having to assist heavily intoxicated musicians in boarding an airplane. Here, he equates the experience with the acquisition of tour management skills, which implies the commonplace nature of such activities and the extent to which TMs must be able to handle them.

> That was just me, a sound guy and thirteen musicians and no money and a lot of booze. That's where I learnt a lot of my tour management skills, it was like trying to herd cats. Drunk cats, carrying musicians onto a domestic internal flight in such a way that they appear to be awake and sober, marched on past, boarding card, sit down, got away with it.

Musicians' age and the extent of their touring experience can be a factor. Younger, less experienced musicians tend to need more oversight. The reason for this is connected, in part, to the backgrounds and life circumstances of musicians. "Young people who are in a band, usually come from a background of they've not had money and they've been working shitty jobs just to make this band thing work and they have to watch every penny" (Joe 2019). Musicians have a long history of having to cobble together a living from multiple types of work and of being associated with a lower social status (Attali 1985, Ehrlich 1985). Relative success, upward social mobility and the practices of touring create a change, sometimes suddenly, that introduces new and unfamiliar circumstances into their lives and likewise has an effect on them.

> ...young bands who just start to make it have many different shiny baubles and lights [that] are distracting. They are suddenly in this world, maybe they've been working really hard as a nobody band for years and suddenly they're touring and people are paying them attention and they're having a good time and they're going to places they've never been to before, some of them have never been out of the country before and it's like woah ... [their] limits have been taken away ... they don't get the chance to necessarily ... travel the world in the sort of style they suddenly get access to, so you throw these people into a situation where they're suddenly successful ... they can go partying, they can get a bit debauched. It often takes them a long time, and certainly some of the artists I've worked with never necessarily learned it, or learnt it quite late in life when they're health started deteriorating that they need some self-imposed limits in order to run it properly as a business.
>
> *(Joe 2018, 2019)*

This situation is not unique to the process of touring. Musicians working in recording studios have been subject to the same concerns around distraction. In the 1970s and 1980s, residential recording studios, facilities situated in remote locations in which artists lived temporarily while recording an album, were deemed particularly attractive settings because musicians could focus rather than be distracted by urban life (see Kielich 2015). The efforts that TMs make to familiarise and acclimatise musicians to the realities of touring, particularly during the early stage of a career, means that TMs are integral to helping artists learn how to tour.

In the same way, older, more experienced musicians need less oversight. Joe (2018) demonstrated the differences, and implied the effect on the TM's role, when he expressed being "lucky" to recently work with an artist who has been touring for several decades.

> They understand that 'hurry up and wait' is touring. So with this lot, there's very little cat-herding as long as you keep the information flowing to them, as long as they know what's going on. They show up on time, they get in the van, they get out of the van, you know. There's an understanding of the routine as well, they like a certain routine, for example.

A major difference is in the artist's response to the TM's communication and the degree to which they follow directives. Instead of needing to be reminded, they simply need to have the information made available to them. The artist's understanding of the nature and realities of touring and their preference for adhering to a routine show that experience can shape their practices and the TM's experience.

A tour's itinerary means that TMs are under pressure to keep to schedule. They are in charge of ensuring on-time arrivals and departures and maintaining awareness of the amount of time spent during stops on the road. Adhering to the schedule can become an "important source of strain" between TMs and musicians and a source of stress for the former (Cohen 1985: 12). Otherwise routine and mundane tasks can become time consuming and ultimately frustrating for TMs. They also reinforce the time constraint imposed by the itinerary and capture the pressure that TMs' experience in adhering to it. Joe (2018) described an example of stopping for gas while traveling on tour:

> If you stop for gas everybody has to think about 'right, do I have to go to the loo or do I need something to eat?' get out and go for it ... this happens especially with young bands, somebody goes to the toilet, gets a sausage and comes back and then somebody else goes 'oh, that looks nice' and then they come back and its twenty minutes later and 'actually I need the toilet' and before you know it you've spent an hour in a gas station, which could

> have been 10 minutes, so your blood pressure tends to go up as a tour manager, fucking get it together. That writ large is everything you know.

He elaborated that musicians gradually come to realise the inconvenience this creates when they see the effects of such delays and will modify their practices.

> Once the first guy starts realising he's been sitting in the van for 45 minutes after he's been to the toilet they start to get it together, they stop being late for lobby calling because it becomes less stressful at the airport, where you have to hurry and check in. They start trusting you as a tour manager.

Making stops during a tour, such as these examples, also create conditions of possibility for band and crew members to get left behind – a reality that Kim Hawes (2019: 90–91) described as "worryingly easy." TMs make choices and engage in practices either in response or to prevent these problems. Hawes recounted a crew member being left at a service station and, upon discovery, the TM decided he would have to find his own way back rather than have the bus turn around. In contrast, to prevent this from happening, Hawes, after becoming a TM herself, made a concerted effort to be aware of who was and was not on the bus when stopping and ensured that everyone had returned prior to departure. In 2017, Richard Colburn, the drummer of the band Belle and Sebastian, was left at a Walmart in North Dakota. Lead singer Stuart Murdoch explained that as he was leaving the store, Colburn was walking in and was left behind – in his pajamas and without a passport or phone – because the "band had become 'blasé' with their system for checking onto the bus" (Beaumont-Thomas 2017). The rest of the band went to sleep and later arrived for their next show in Minnesota and realised he was missing. They devised a new system of placing a sign in the bus driver's seat when someone leaves the bus at night in order to prevent similar incidents from occurring in the future (ibid.).

The significance of adhering to an itinerary and the importance of the "instrumental component" of the TM's responsibility in ensuring the smooth progression of a tour are further evident when musicians defy the latter's communications, and their actions impact a live music event. Such a situation also reiterates the complicated power dynamic between the two parties. A local production manager recounted a relevant incident. During a day off, a musician flew to another city to spend the day with his family though the TM had advised against it. In trying to return for a concert, he encountered several flight delays and was also unable to communicate with the touring party because he forgot his phone. He arrived after his own concert had already started and walked onstage during the performance. The efforts of the TM are intended to prevent such problems.

The specific nature of touring, and the manner in which it shapes and affects everyday life, can become a determining factor in organising an itinerary. It can involve making adjustments in order to meet the needs of specific musicians.

> The artist I'm working with at the moment can't sleep, very bad at sleeping on buses, so what I have to do is try and make the drives as short as possible or day drives or fly when possible. Anything up to 4:00 in the morning it'll be, aftershow, drive, anything up to 4:00 in the morning, it'll be get off the bus and check in to the hotel at the next place, which kills everybody else. It's really bad for me, once I get off the bus and check in to the hotel, I can't go back to sleep again. You know, I don't sleep properly for days at a time. But [the artist will] check in at 4:00 in the morning, sleep till noon and … be fine.
>
> *(Joe 2019)*

TM Ben Price also highlighted an artist's sleep schedule as a significant factor in planning an itinerary based on how the pace of "travelling inevitably means having little time to sleep" (Raven 2015). His approach is that "when booking travel you have to plan to allow [the artist] the most amount of rest possible." In this case, the artist being well rested was also of particular importance in order to ensure that he was able to look after his newborn after returning home (ibid.). These examples demonstrate that TMs arrange travel itineraries in order to accommodate the artist's needs and preferences. It also shows that doing so ultimately benefits the artist while, in the first case, negatively affecting the rest of the touring party. Making such an adjustment for the artist disrupts the lives of the people who work to make musicians comfortable. However, such a trade-off is deemed necessary in order to realise live music events. This activity highlights that being support personnel means foregrounding the interests and meeting the needs of others while those of the TM are compromised and their status as secondary is reinforced.

Cohen's (1985: 12) second component of the tour guide's role involves "responsibility for the cohesion and morale of the touring party." TMs, along with PMs, are involved in making sure the relations between the various members of the touring party work properly and problems that arise are handled. Findings about TMs deviate slightly from Cohen's framework here, however. The extent to which TMs work to keep musicians in "good humor and in high morale" varies (ibid.). Perhaps more precise in terms of the notion of cohesion in the touring context is that TMs should "bring the best out of their team" (Hawes 2019: 348). To do so, TMs learn to interact with and manage the various personalities of the touring party. As has been demonstrated in this book, this involves "taking an interest and concern in each person, knowing when someone needs an encouraging word, when they need a kick in the pants and when they just need leaving alone. It involves letting them know you appreciate them" (ibid.). TMs are concerned to "keep them happy" by meeting their needs, working according to their preferences and managing their expectations when circumstances fall short rather than through the performative tactics or emotional labour utilised by a tour guide. They do not need to agree with artists' particular needs but must also not dismiss them. Rather, they need to take them seriously and recognise them as an essential component to an artist's comfort; the particular reasons such needs exist may emerge during the tour (Atkins 2007: 253). TMs also try to "prevent the emergence of tensions" between musicians, recognise when problems may arise and manage conflicts that occur (Cohen 1985: 12). Doing so strongly depends on their ability to get to know and understand the musicians for whom they work.

Third is the interactional component, which has to do with the ways in which the TM operates as an intermediary between musicians and the local area. These responsibilities include acquiring necessary services and amenities during the tour, which involves working with the local population (i.e., the local crew). This was addressed in Chapter 3 in regard to the rider and aftershow food, and the general navigation of the venue and backstage area. Operating as an intermediary also involves integrating the artist into a local setting as well as insulating them from it (13; Schmidt 1979: 454). Doing so is accomplished by positioning themselves between the group and the environment to make it non-threatening (13). During participant observation, TMs demonstrated this aspect of their roles in relation to the availability and features of artists' dressing rooms.

Upon arrival at the venue, usually between 8:00 a.m. and 10:00 a.m., the TM (or PM if onsite first) will meet with a member of the local crew, usually a member of the hospitality staff, and do a "walk through" to assess and assign the available space. Rooms are assigned based on a combination of the venue's particular available space and the number and needs of the artist and larger touring party. A touring party's spatial requirements are specified in the rider. A venue may easily accommodate an all-male four-piece rock band with extra rooms available for hospitality, a production office and the crew. That same venue may present challenges for a six-member R&B group that features a principal or star performer, a backing band and female backup singers. Spatial arrangements, and the efforts put into them, are indicative of the dominant factor of status in the division of labour, both in the entire touring party and within the individual musical group.

As explained in the Introduction, musicians may be "particular sorts of workers" who are "people seeking to do jobs," but in the backstage environment, their status as "performers, celebrities, and stars" (Williamson and Cloonan 2016: 8, 10) is central. Dressing rooms tend to be prioritised by the individual status of members of a group, role in the group and by gender. It is based, in essence, first on the people deemed most important to the performance (e.g., the lead singer, who is also a famous rock icon) and then by gender. Lead or principal/star singers, male or female, are customarily given a private dressing room. If several members of the band have similar status (e.g., they are both primary instrumentalists and comparably famous), they will usually each request a private room in the rider, but the available space may not be sufficient. The remaining members who constitute "the band," or the group of musicians who back the lead singer, often share a room, but genders will not be mixed.

Existent studies have indicated that the backstage conditions women encounter on tour are inhospitable and lack privacy (Bayton 1998: 133–134, Kearney 2017: 159–160). Observations contrast significantly with those research findings. Women who are considered part of the primary group of performers, who may or may not have the same status as the principal, are given a private space, though two female backup singers may be allocated a shared room. The privacy accorded women backstage is one of the rare instances where their exceptional place in the male-dominated nature of touring and the music industries more generally works in their favour. Dressing rooms also represent and reinforce the power and respect that women of status occupy in the music industries. In one case at Venue A, male band members were lounging in the dressing room of a prominent female singer and upon being told that it had been assigned to her, they made panicked exclamations, quickly jumped out of their seats and fled the room.

Dressing room space creates consistency and stability for musicians within the continuous mobility of touring. Ensuring these rooms are suitable and comfortable is likewise an important part of the daily tasks for TMs. Ideally, dressing rooms and green rooms offer musicians privacy and space for relaxation, socialising and preparing for performance. The quality, features and ambiance of dressing rooms vary drastically, however, and the multipurpose nature of venues and quick turnover of concerts mean their facilities may fall short and feature inadequacies. One touring party carried their own amenities, and the TM recreated a "home environment" by temporarily installing turntables and games in the dressing room. Others converted the dressing rooms into rehearsal spaces for artists. More commonly, however, once a venue's available space is assessed and rooms are assigned, TMs will delegate tasks, if and as needed, to the local crew to make any necessary adjustments to the room in time for the artist's arrival.

TMs may encounter challenges if the needs of the artist are at odds with the character or resources of the local environment, but they are also in a position to negotiate and manage expectations. The earlier the rooms are assessed and assigned the less of a time constraint is involved in addressing and troubleshooting issues. These requests can be as simple as removing an unused cooler from a dressing room to create more space or as impossible as eliminating the noise emitted from a building's internal ventilation system. TMs pay attention to such details, and make adjustments accordingly, because they can affect not only the artist's comfort during the day but also the longer trajectory of a tour. Issues such as air conditioning in a dressing room are monitored for the adverse effect they can have on a singer. At Venue B, the TM explained that the excessive use of air conditioning at some venues during a period of extreme heat had gradually affected the singer's voice and resulted in the need to cancel a show.

This aspect of a TMs job closely resembles that of a producer's role in a recording studio. Negus ([1992] 2011: 84) states that producers must create conditions in a recording studio "which are most conducive to the working style of a particular act." They use several techniques, ranging from "subtle psychological and social skills involved in creating a dialogue and a repartee between those involved to physically creating a particular environment" (ibid., Stokes 1976). This directly reflects the selection, and adjustment, of the most suitable rooms for individual members and the delegation of tasks to the local crew to modify them as needed. In relation to the touring world, however, Negus's view that producers "create" conditions should be nuanced. Creating particular conditions can suggest novelty, whereas what TMs are doing is trying to prevent potential issues or remove noticeable differences. They want space that is consistent and comfortable so that artists know what to expect. A local production manager with 15 years of experience in the live music industry, captured the difference. He expressed perplexity at the "obsession" he observes that touring crew have with *controlling* their environment. His analysis is a useful means to convey the importance of maintaining consistency and that this is the ultimate purpose of the TM's actions when assessing and delegating dressing rooms. Creating consistency, or controlling the environment, is a means to construct and maintain everyday life.

The selection of rooms, and how to adjust their particularities for the duration of the day, is one of the key moments when a TM's knowledge and understanding of an artist are crucial. TMs must anticipate the needs of musicians in view of the character of the room. Their ability to do this effectively can come from years of working with the same artist and clearly understanding the distinct personalities in a group. TMs use different approaches to determining suitable rooms and demonstrate variation in the confidence of their choices. One TM, in charge of a large touring party with multiple artists, took the realistic view that the number of people to please means it cannot be done all the time. Her method was to look at the various dressing rooms and imagine how the different acts would react to each room. Another changed his mind several times, prior to the artist's arrival, about where to put the various group members. He was particularly concerned about the wardrobe person who he believed would "freak out" about the ventilation noise in her proposed room and needed a solution to prevent this.

Some TMs are less familiar with or completely misjudge an artist's preferences. Reflecting the evaluative criteria of the local crew, a local manager observed that the character of a day can be completely changed in such circumstances; an artist can be "easy," but problems occur if the TM is not capable. This was evident on a show day with a group consisting of two principal women and two men, one of whom is also a lead vocalist. They requested three dressing rooms plus a crew/hospitality room; however, the layout of the venue meant that these arrangements could not fully be met. The TM was filling in for another tour manager and arrived at the venue 90 minutes later than planned, which generated a time constraint. She initially decided to put the primary woman principal in Dressing Room 1 and to divide Dressing Room 2 between the men and the second woman. Doing so meant that the local crew and the promoter had to create a physical divider in the room, but this could only be fabricated with six-foot moveable dividers. The result failed to create a truly private space, which the TM was adamant was necessary for the second woman and her assistant. She also casually yet confidently stated that the men in the group would not mind which room they were in because they would simply adapt to the space they were given. Upon arrival in Dressing Room 2, the male principal, in contradiction to the TMs interpretation, was dismayed at having to share space with the second woman principal, especially in such a makeshift manner. He exclaimed that "this sucks" and expressed confusion as to whether he was in the correct room or the crew room. His association of inadequate space with the crew suggested a sense of hierarchy. The TM was noticeably shaken by his unexpected reaction and changes were quickly improvised and coordinated with the promoter and the local crew.

The manner in which space is handled seems to be affected by the relative popularity of an artist. Some performers were older, very well-established artists in their 60s and 70s who have continuously worked, remained household names and achieved a certain iconic status while experiencing fluctuating popularity across a career. Others were musicians with careers of more than 20 years who were most closely associated with a short period of perhaps one or two album cycles during which they were most well-known and are best remembered for those few songs. TM Mark Workman (2012: 314) explained in his book that one of the most difficult jobs a TM can have is "working for a veteran band that was once much bigger than they are now." He elaborates that the difficulty arises from the fact that such artists have a hard time adjusting to a lower level of touring (ibid., see also Webster 2011: 131).

During fieldwork, several artists were bands that experienced their peak popularity during the 1990s, but whose success subsequently diminished, and now generally maintain visibility on tours primarily associated with the nostalgia market, some of them as co-headliners with other artists of the same calibre. Their TMs were particularly dissatisfied with the available space, and their complaints and requests could be interpreted as attempts to cope with their boss's diminished musical relevance. Just as the quality of cultural products reflects on the workers that created them (Hesmondhalgh and Baker 2011: 181–199), the popularity and relevance of an artist is intertwined with a TM's reputation. In one instance, the TM insisted that his band was to receive the bigger room and the "bigger everything." Though the tour was billed as co-headlining, he countered that the other group was actually the support band, and that they were not to share space. This was particularly significant due to the fact that the band relegated to support status had once been the most prominent band to come out of the local area. The TM criticised the appearance of the dressing rooms and articulated that he wanted the band to feel comfortable rather than as though they were in a lunch room. After requesting furniture to replace the large table and chairs in Dressing Room 2, he learned that the only available furniture in Dressing Room 1 could not be moved due to the unit being one interconnected, large, heavy piece. In response, he stated that he "gives up" and instructed the local crew to remove some of the tables from Dressing Room 2. He then complained about the walk from the dressing rooms to catering, concluding that it was a personal affront because the closer placement of the facilities' meeting rooms meant that the touring party got "the short end of the stick." The same distance to catering was interpreted much differently by the TM of an influential band whose lead singer is considered an icon. The second TM was concerned about the singer's privacy and inquired about the closing time of the facility's public area in order to determine the best time for the singer to walk there.

In another instance, the TM noticed some discolouring and stains on the sofa in a dressing room and became very concerned about the band seeing them. He requested that the hospitality staff cover the sofa before the band arrived three hours later. To make that happen, a member of the hospitality staff went to a store, purchased a set of sheets and attached them with safety pins to the sofa, which met the approval of the TM. Upon inspection after the show, it was unclear if anyone had actually sat on the sofa or noticed it at all. Enacting power through the management of space gives the illusion that the artist's relevance has continued past its peak by creating a false sense of "star treatment." These TMs use the labour of the local crew to protect the artist, and themselves, from the realities of the artist' career positioning. These instances further indicate the omnipresent influence of status in the working lives of musicians, and how that positioning is negotiated and reflected in the actions of their support staff. Compensating for diminished status requires more effort whereas consistent status is ordinary. In both cases, TMs are aware of the ways that status can or does affect the artists they support and perform their roles accordingly.

Maintaining control over the environment also involves regulating the behaviours of musicians as needed. In the backstage area, the handling of alcohol acquired for the rider, and its placement in dressing rooms, was at times a concern for TMs and other members of the touring party. In some cases, placing alcohol in the dressing room was completely acceptable and not seen as a potential problem, while others did not want any or all of the alcohol to be readily available. In one case, after a dressing room was set by hospitality staff, the TM removed two bottles of wine from the room during the afternoon, saying he did not "want to encourage excess." A road manager asked the local crew to leave all of the alcohol in her production office except beer, which was allowed to be placed in the band room. She kept the principal singer's liquor and wine with her until she determined the suitable time for placement in the dressing room. Regulating alcohol also occurs outside of the dressing room space, and the need to do so does not necessarily diminish with age. The TM for a band, the majority of whose members are in their 70s, requested an additional supply of alcohol to be purchased for transport to the next city. When the local crew returned with the alcohol, the TM and the PM immediately hid it in the bottom of a flight case, burying it beneath other items. The TM stated that if they did not do so, "it will all be gone."

Dealing with alcohol is among the considerations and judgement calls TMs make when attending to musicians' requests, based on the acquired knowledge of their personalities and habits. TMs appear to take these decisions upon themselves, without the input or request of the artist, meaning that they effectively say no to musicians in a non-confrontational manner. This regulation is ultimately based on trust between the two groups, as part of the larger function of the TM's role to attend to the tour and the artist's needs, but it is also a key component of the TM's aim to make the concert happen by preventing an undesirable outcome. These examples also show that the regulation of musicians does not only fall onto the TM, but is a collective effort on the part of the touring party to maintain consistency in the working day.

The interactional component can be additionally applied to how TMs manage relations between fans and musicians in cities across a concert tour. Part of the everyday life of a TM involves interacting with the fans of the musicians for whom they work. This activity also reveals the ways that TMs protect musicians. The majority of this interaction occurs on show days when the artist and TM walk in and out of a venue as they arrive or leave. Arrival at a venue is a carefully coordinated activity, and the logistics of it vary with type and location of venue. The arrangement of an arena, for example, can greatly inhibit fans' access to musicians as buses tend to pull into enclosed areas or inside the venue. The TM will notify the PM, who is already on site, of the impending arrival of the TM and artist to the venue. The PM will help to prepare the stage door by communicating with venue security and, if possible or necessary, ensuring that there is a barrier set up to facilitate smooth entry to the venue for the artist. When artists have offstage contact with fans, TMs oversee and manage the interactions between them. As TM Tom Begley told *The Guardian*, "most bands want to meet the fans" (see also Baym 2018) and for the TM, it is a "juggling act between allowing some access and autographs and ensuring privacy when they want it" (Simpson 2015). These types of situations usually include autograph signing and occur because the artist wants to and has agreed to do so. Artists' willingness to sign autographs, and the amount of time they spend doing so, varies enormously.

> Meeting 500 screaming fans certainly isn't for everyone. I've had musicians say, 'I don't want to take 700 selfies with the fans. I've just put on my pyjamas.' On the other hand, whatever needs to be done is the tour manager's job. If the singer wants to stand in the cold for an hour and a half signing autographs, I'll stand there with them.
>
> *(Tom Begley in Simpson 2015)*

Signings tend to occur upon arrival and/or departure from a venue, at the stage door, and are often mediated with the assistance of venue security, who are briefed in advance by the TM about an artist's signing practices, and the physical barrier that is set up between musicians and fans, though there are exceptions to this arrangement. This is typically a relatively straightforward process during which time fans get autographs and TMs oversee and help manage the interaction between them and the artist. They also try and prevent situations that are at odds with the artists' wishes. These include excessive requests for autographs from the same person, or "autograph hounds" or eBayers who are not fans and are only interested in autographed materials in order to sell them.

Fans seeking autographs can create unintended problems in the working day of a TM. At one venue, a fan mailed a guitar pickguard to be signed by the band's frontperson, who agreed to the request. However, while the sender included a return address, there was no pre-paid return envelope. The TM was annoyed and frustrated by this as it put the touring party in the position of having to take the time and spend money to return it to the sender. The TM explained that had they included a pre-paid envelope, it would not be a problem.[2] Such small instances create unanticipated and additional tasks for TMs that become further complicated because they require the resources of a local area that may not be familiar.

TMs also function as gatekeepers between fans and musicians. They are commonly given gifts and letters from fans to deliver to artists. The handling of such items, consistent with many aspects of the touring world, varies greatly. Research findings for this study suggest that many times such items that are given to TMs do not actually reach musicians. The TM's role as a mediator in this sense can be seen in how fans will give gifts and letters directly to musicians when possible, but musicians may not receive them when given to the TM. Joe (2018) stated that the TM may discuss having received items intended for the artist with them, and the musician may be compliant in not receiving them. In some cases, the artist is simply uncomfortable with or embarrassed by such gifts and letters, while other artists do welcome the receipt of such items while on tour.

TMs at times exhibit a sense of reluctance or awkwardness at handling such items and fan encounters, which is based more on their own reactions and feelings about a given situation than it is to do with managing artist's preferences and expectations. Part of this has to do with the nature of some of the letters and encounters. Joe (2018) described his experience:

> ...every night when you hear about my sister's got leukemia it gets pretty depressing. It's a constant thing ... I've got somebody with a fatal illness that needs a request from the stage or would like to meet [the artist] or, just all the time ... With every artist I work for. So you start to be a bit hard-hearted about it, ignore all of it ... Sometimes you get a picture of them with a drip and a ventilator. It can be quite upsetting.

In some cases, TMs protect the artist from such interactions with or correspondence from fans. In contrast, Nancy Baym (2018: 42–43) indicated that musicians encounter "all sorts of bleeding heart situations" through correspondence and meetings and take "particular joy in feeling they had soothed people who were suffering from illnesses." This further supports how the practices of touring and its management vary according to artist preferences. However, the practice of gatekeeping for TMs means confronting difficult and delicate situations. TMs risk exposure to sensitive subject matter with which they are not trained to deal with and which can be a source of stress. This shows that the kinds of activities TMs encounter in an effort to do their jobs also come with effects, and that the act of supporting another person has the potential to put them in situations they find uncomfortable. Such occurrences indicate the extent of the support given to artists while potentially leaving TMs vulnerable. In doing so, status and hierarchical differentials between musicians and TMs are reinforced and expressed by who gets protected and who does the protecting. While many TMs, based on their own experiences with music during their youth, "understand the feeling" (Begley in Simpson 2015) of fans wanting to meet or have some type of contact with musicians, Joe's quote also reveals that TMs learn to protect themselves as a result of situations that surpass their limits and comfort levels by developing a detached approach to interacting with fans.

During participant observation, a veteran TM also expressed reluctance about having to engage with fan correspondence. He summarised his feelings by scrunching up his face, as if cringing, as he made a comment about "when you open something from fans." However, their frame of mind about interacting with fans also relates to the kinds of personal relationships and familiarity that develops over years of working with the same people. Such familiarity creates a normalcy that greatly differs from the special status accorded musicians by fans, and that is maintained by distance and anonymity. This further indicates that an artist's status, and their differences with others, is a normal part of the daily working life on tour.

The uncertain nature of fan-artist interaction that TMs encounter has a longer history. As an example, the former TM for The Animals, Tappy Wright (2009), recounted a 1964 incident at the Ed Sullivan Theater in New York. As the band and touring party entered the venue, a girl had three of her fingers severed after being slammed in the backdoor entrance to the theatre as she tried to meet the band. The girl was brought into the building and Wright and the band tried to comfort her while the drummer called an ambulance. Wright noted that she seemed more attentive to having met the band than to her bleeding hand (23–24). While this is an extreme example, it demonstrates a breadth of the spectrum of kinds of situations which TMs encounter and must attend to when working with artists.

TMs are also involved in facilitating pleasant and enjoyable situations for fans. Requests are made, and sometimes granted, for onstage marriage proposals, and some artists permit and encourage fans to dance on the stage during parts of the concert. In these cases, TMs are involved in making necessary arrangements with venue security, arranging a route for fans to access the stage and ensuring safety during the activities. TMs also work to ensure the safety of artists who leave the stage to interact with the audience during a show. At Venue A, one artist regularly walked through the crowd, playing guitar and singing along with fans. At times, he would stand on top of chairs or the dividers that separated sections of the audience. Throughout this process, the TM followed him and watched him closely. In these circumstances, TMs take on a role more closely associated with security, and such acts quite literally reinforce the centrality of looking after artists to their roles.

At Venue A, after show VIP meet-and-greet events were another site of interaction between TMs and fans. These events are part of a higher priced ticket category that usually includes preferred seating along with the opportunity to meet the artist. These events are in distinction to guests or fans that may receive aftershow passes from the artist or TM. VIP meet-and-greets that are built into a ticket are paid for, whereas aftershow passes are given out for free.

The VIP events were held in the employee lounge next to the dressing rooms. TMs are responsible for explaining to the group of fans, prior to entry to the backstage area, what will happen and what to expect during the event, and then escorting them to the designated area. During the event, the TMs stayed in the room and quietly monitored the activities.

One TM recounted problems that occurred with fans following paid VIP meet-and-greets on a solo tour of a member of the group he has long worked for. The events had highly limited attendance numbers and, to facilitate ease and address any issues that may arise on the day of the event, the TM gave his cell phone number to each of them and directly asked them to delete it afterwards. The next time that the full band was on tour, some of the fans who attended the meet and greets on the solo tour kept the TM's number against his wishes and called to ask if they could get a backstage pass for after the show. The conventions around these VIP meet and greets, and the fans' actions, further position the TM as a gatekeeper. However, the fans' blatant disregard of the TM's instructions suggests that they see tour managers solely as a means to an end based on their access and proximity to an artist, and attempt to take advantage of them on the basis of their roles. This example highlights the conflict between authority and subordination, and the notion of support as secondary is reinforced.

The fourth and final element of Cohen's major components of the tour guide's role is the communicative component, and it is the most important. The "kernel" of the TM's responsibilities is to provide accurate, up-to-date and specific information to musicians and the wider touring party (Cohen 1985: 15). For tour guides, providing information involves giving informed knowledge about the local area being visited, and also interpreting it for tourists. For TMs, this component is much more functional in nature and aimed at the overall goal of realising live music events. They need to be knowledgeable about the logistics of travel rather than the area. The communicative component includes providing information about arriving and departing from all relevant locations on a tour, such as hotels ("lobby calls"), transportation ("bus calls") and venues, and communicating details about the schedule of show days. Tour management is all "about the communication and keeping people aware" (Joe 2019). The process works because the TM sets the itineraries, communicates them and provides reminders when necessary. Doing so successfully also requires that musicians are punctual and "get into the right habits."

Adhering to an itinerary and the need to coordinate a group of people involves careful planning in order to ensure travel is smooth and on-time. One illustrative example is arrival times at airports. Traveling to the airport can mean going as a group or, such as at the start of a tour, coordinating the arrival of musicians from different locations at the same time. The crew may be responsible for getting themselves to the airport, whereas the TM will arrange ground transport for the members of the group. If the TM does not allow for ample time, any type of accident on the way or problem at the airport could result in missing a flight which has the potential to result in missing a show. Some musicians are "furious" if they have to arrive at the airport too early and others "know that two hours before your flight is perfectly acceptable and indeed a great idea" due to the way that it creates a time buffer (Joe 2019). Stress and time pressures are also reduced by traveling the day before a show, and by avoiding taking the last flight out. If musicians have press engagements, the TM will handle the arrangements to make sure they get to and from radio stations or other locations at the appropriate times, and may or may not accompany them depending on the artist.

The TM makes decisions about travel, accommodations and other necessities based on the personal needs and preferences of musicians. Joe (2019) described the process of informing the artist he works for of the daily schedule.

> ...[artist] [is] quite technophobic or at least she doesn't want other stuff on her phone, so I've got to copy and paste everything ... and put it in her iCal. So I've just got to adapt, so that information flow I've got to really make sure she knows what's going on, but she's really good at it, she'll get something and read it and everybody gets it emailed as well ... and usually within two minutes of reading it she'll come back with any questions she's got, so she does pay attention to it and does want to know.

This example also shows the artist-specific nature of the role and that TMs adapt based on their preferences. Artists generally have preferences regarding flight times and when they arrive somewhere in relation to the start of their work responsibilities. TMs take any choice in the matter away and "book the flights that they like. You don't even have to think about it, you just send them some flight details ... once you know their preferences" (Joe 2019). The same practice applies to hotels. In her autobiography, TM Kim Hawes (2019: 206) recalled having to alter hotel reservations for guitarist Marc Moreland, who did not like elevators. She had to ensure his room was always on the ground floor or find another hotel that had such availability.

The TM makes arrangements and relays the information via the day sheet the day before or in the tour itinerary. The communication of relevant information also has the ability to be challenging for TMs when musicians fail to comply. The conflict in their roles can be further observed as the question emerges as to which party is actually in charge.

> ...[musicians] don't always listen either, you say to them the night before [that] lobby call is at 9:00 a.m. and you've given them an email and you

> might even have put it on a paper and shoved it under their hotel door, but sometimes it just doesn't occur to them to think about it until ... And they won't do anything till you ring their phone. 'We're leaving in half an hour,' you know, or 'we're leaving ten minutes ago, where are ya?' Then it's a bad habit to get into.
>
> *(Joe 2018)*

Kim Hawes (2019: 292) described an incident in which she had provided a wake-up call to a horn player who subsequently did not meet the rest of the touring party in the hotel lobby. Hawes called again to inquire, and the musician said he was in his room, to which Hawes asked him to hurry downstairs as they were almost late. After he did not appear, Hawes went to his room and knocked on the door. The horn player eventually answered and stated that he'd gone back to sleep. Hawes instructed him to get dressed and go to the bus, which he complied with.

Ken Barr (2009: 51) noted a similar pattern amongst musicians during his career as a backline technician and stated that "some of the most talented people in the world cannot get the concept of time, schedules, being on time." In turn, the TM is "constantly having to remind them of what their routine should be" (Joe 2018). Hawes (2019: 212–213) gave an example of this practice in her autobiography. She recounted an incident in which the guitarist for Concrete Blonde, Jim Mankey, knocked on her hotel door at 10:00 a.m., and woke her up, complaining that he had not received a wake-up call and feared he was late. When she asked if Mankey had seen the schedule she left under his, and everyone else's, hotel room doors that stated the touring party would not be leaving till 3:00 p.m., it was clear he had not consulted the document. Mankey had already checked out of the hotel and Hawes had to ask reception to return his room key, which they agreed to do. Hawes noted that after this incident, Mankey always followed his schedule (214).

The practice of reminding artists of their schedules has effects. Such communication creates a "bubble" around musicians.

> What day is it? Where am I? You know because you can just let go and someone else will take care of absolutely everything, they'll hand that boarding pass to you, they'll, you're off stage they'll say get into that car, you're going to that hotel, here's your hotel room key, you know. You need to be at the lobby at 9:00, if you're not in the lobby at 9:00 you'll get a phone call getting you down to the lobby at 9:00, you know. There's no independent thought needed should you choose to be that way. And some people like that cause then they come alive at five to nine when they walk on stage.
>
> *(Joe 2017)*

While the TMs' role in alleviating musicians from "normal household tasks" is intended to facilitate focus on live performance, it can also have unintended effects. Frith et al. (2019: 176) observed that a tour's itinerary is indicative of "the complex logistics of a corporate tour" but that it also "hints at the way successful rock musicians were infantilised" (ibid.). They mention the day sheet given to members of the touring party that provides information about the city, date and schedule. It is often distributed by being placed under hotel room doors and "was laid out with great clarity." Their emphasis suggests that this is another component of infantilisation. In and of themselves, documents intended to communicate important information about the plan and organisation of a tour to the entire touring party, insulting in their extreme clarity as they may be, are not infantilising. Keeping track of a schedule is an essential component to the success of a tour. Rather, it is the repetition and extent of this practice that can be understood as infantilising. What the itinerary and day sheet represent are part of the wider effects of handling tasks for musicians and reminding them of their responsibilities. Road crew members are expected to keep themselves informed and should not expect such reminders. They will "get barked at" and cause the TM to be "extremely annoyed" if they are not aware of information that has been communicated via the day sheet because it is "their responsibility to check that" (Joe 2019). Musicians become "used to having somebody around them fixing everything for them" (ibid.). Wendy Fonarow (2006: 136) made similar observations about the more generalised practice of support or assistance provided to musicians by members of the touring party.

> ...musicians learn to not attempt to solve problems at shows. Most successful performers are infantilized at gigs from their experiences of being on tour, where all matters are dealt with by tour manager and crew. Most musicians have the expectation that if they just stand there and do nothing, their problem will be solved by other industry professionals—and it usually is.

TM Kim Hawes recounted several relevant examples in her autobiography. The "expectations bands have that a tour manager is there to solve their problems" was made explicit by a badge she received from a group stating "Kim'll Fix It" (Hawes 2019: 347–348). While this was likely done in jest, and an expression of the workplace humour that circulates on tour intended to create camaraderie, it is also a reflection of the perception of TMs. She recalled encountering particular issues with airline meals for bands, calling them a "great source of trouble" (291). Part of it she attributed to the kinds of requests that were made by musicians, in her observation, that she knew were unlikely to be met, and which she believed the artists made "for the hell of it" (ibid.).

If requests could not be met, it was her responsibility to both "placate" the artist and work with the airline crew to find out what they could provide as an acceptable substitute (ibid.).

In other instance, one of the band members and a crew member got drunk at an airport after checking in and subsequently lost their passports. Hawes (2023: 195–196) had to convince an airport customs officer to make an exception for the band member and crew member and allow them through. She did this by locating a record shop in the airport, purchasing one of the band's albums that included a photo of the band member in question, providing paperwork proving their identities and vouching for the crew member who had no other form of photo identification. The officer did make an exception, and Hawes stated she would conduct her own passport checks with artists going forward (196). Hawes also had to conjure up excuses for artist's behaviour to avoid problems. She recounted an incident in a town in Italy in which two band members decided to remove all of the Bibles from the small hotel where they were staying, and then throw them into the street (199). When the outraged manager and promoter found out, Hawes told them that the band did not mean to do it; rather, she explained that they had found a box in their room that did not belong to them and decided to get rid of it without realising its contents, which they would have never thrown out had they been aware. She explained that they should have contacted the hotel staff to handle it rather than doing it themselves. Hawes was going to suggest that the band pick up the Bibles, but this had already been done by people in the town (ibid.).

Findings support the idea that TMs engage in practices that can have enabling effects, and that musicians come to depend on them for a variety of tasks. One TM explained that he is the person who reminds the band members of people they are supposed to call in a particular city. However, if they do not make the calls, he does it for them. TM Iain Williamson told *The Guardian* that "bands do turn into children – you have to make sure they're all on the bus and all in the hotel" (Simpson 2010). In another instance, observing a TM discuss lunch options with an artist in the dressing room was striking in that the interaction seemed to position the latter as a child. The musician, in her early 70s, sat on a sofa with her hands tucked under her legs as the TM read selections from a menu to her. However, the same musician often personally calls the hospitality staff to discuss available food options in the local area. In his book about touring, Martin Atkins (2007: 253) acknowledged a tension between understanding and accommodating the needs of artists and treating them like children. He used the example of his youngest son needing a specific stuffed animal and pacifier in order go to sleep, and that he provides them without question, as a way to illustrate how TMs should accommodate the needs of musicians. The "trick" is to do so without "treating the artist like a two-year-old" and to avoid disrespecting an artist's "pre-show set-up psychology" (ibid.). The choice of example strongly reinforces notions about infantilisation and also suggests a particular view of musicians. It also implies that TMs are aware – or should be – of the potential implications of their working relationship and work to avoid such treatment through sensitivity to the nature of their interactions. While dependency can develop, it can also vary and stop. Effort and awareness on the part of the TM can play a role in deterring dependency, and musicians may naturally outgrow this aspect of the relationship.

> ... I've seen people go through phases of it and then kind of shift, especially maybe a new kind of session musician in a band will suddenly get really reliant on me and 'oh Joe here's my washing' and 'oh Joe can you do this and can you do that' and then I'll enable it and then gently rebuff them if and I think most people, after awhile, start to not like that relationship anyway and become more independent themselves. You know, days off they're not going 'Joe, what can I do and can you get me a cab' it's just I just don't hear from them and they don't want to hear from me either.
>
> *(Joe 2019)*

This type of dependency can become a habit that extends and continues off the road. Joe (2018) recounted a TM colleague who would receive phone calls from an artist four months after the end of a tour to ask if the TM knew where the artist's glasses were. After leaving touring, Kim Hawes (2019: 337) described becoming aware of "how dependent some of [the people she had worked with] had been on me for the handling the world of everyday." Her reflections do not explicitly or exclusively refer solely to musicians, but they demonstrate the significance of the TM's role and responsibilities and the long-term effects. Similar to how, for many years, Hawes had told them "when they would be getting up, where they would be going, how they were to get there" she then experienced, after moving from England to a state closer in proximity to LA, being "asked what to do about faulty plumbing and broken washing machines" (337–338). Being close by was a primary factor in this. Her closer location to the friends she had made while working on tour created a sense that she "might be able to help" because she was "just across the state border." (ibid.). Due to her convenient geographic location, she was perceived as an obvious source of assistance based on the nature of her work role. This mimics, and thereby highlights the significance of, the connection between a TMs proximity to an artist and the support they provide on tour, the comfort and stability of which can be reproduced in the minds of musicians long after the working relationship ends.

Such activities, and the related infantilisation of musicians, are part of a wider set of conventions in the music industries of which touring is only one component. Outside of live music, musicians may develop an emotional investment and come to rely on their artist managers (see Jones 2012: 80) who can fulfil tasks that resemble those of a personal assistant and who have a payment arrangement similar to that of TMs. Keith Negus ([1992] 2011) highlighted that support personnel at record labels are involved in similar practices that can be linked to the same effects. He described how publicity officers

> must respond to the needs of the various acts who are signed to the label or handled by the company. This can involve preparing artists for interviews, photo sessions and performances; giving them morning calls, making sure they turn up where and when they are supposed to; booking them into hotels and health clubs; arranging transport and generally being at their beck and call.
>
> *(117)*

This shows that other types of workers take on the same responsibilities as TMs in contexts outside of live music, and musicians are regularly surrounded by people who attend to tasks on their behalf. These activities are also an extension of musicians' status as artists and stars and the associated treatment they have been deemed worthy to receive. The following section will analyse the term "babysitter" commonly applied to the TM.

Being a Babysitter

As I have discussed elsewhere (Kielich 2021), TMs are often referred to colloquially or facetiously as "babysitters" due to the nature of their working relationships with musicians. During participant observation, a TM stated, in reference to his career as a tour manager, that he had always been a "babysitter." In his guidebook about tour managing, former TM Mark Workman (2012: 312) noted that his ex-wife used to refer to him as "the highest-paid babysitter in the world." As "babysitters" TMs' related work activity is likewise called "babysitting."

Generally speaking, babysitting refers to the act of providing nonfamilial childcare "on a temporary basis for pay, typically in the home of their employer" (Forman-Brunell 2009: 15). The use of the term problematically equates musicians with children and insinuates that, as adults, they require the supervision of other adults in the same way that a child does. It aligns the TM's role with a type of labour commonly associated with female teenagers and undermines the skillset and its significance in the realisation of live music. The job of a babysitter does have characteristics that support reasons for the term's use in relation to TMs. The analogy mirrors the work arrangement between the two parties as TMs work for the duration of the tour, and do so in close proximity and shared living space. A babysitter may fulfil the role temporarily, but can also do so routinely and consistently over a long period of time, just as tours are short-term but the same TM may repeatedly occupy the role for an artist. The use of the term "babysitter" also reveals the nuanced power relations between them. Babysitters are usually paid by a parent, meaning a third party is hired to look after someone younger who requires supervision. In contrast, musicians directly pay TMs to look after them. A babysitter, for the duration of their job, is considered the authority figure and is likewise granted a degree of power in the relationship, which suggests that TMs are also granted authority over musicians as needed. TMs do exercise some power and authority, but as indicated, the limits can also be unclear and difficult to establish. The similarity is further evident in the ways musicians will disregard the communications provided by TMs. This mirrors those situations in which children being babysat may perceive the babysitter as less of an authority than their parents, and therefore less threatening, and as facilitating greater opportunity to "get away" with inappropriate behaviour.

The term "babysitter" draws attention to the ways that the notion and activity of care factors into the dynamics of the working relationship between TMs and musicians and plays a part in shaping the character of the former's role. Babysitting is defined by the act of childcare, and the idea of "mothering" is the "paradigmatic act of caring" (Tronto 1993: 109). As articulated in Chapter 1, care is a concern for, or action directed towards, others. It suggests a "reaching out to something other than the self" and positions another person's needs as the "starting point for what must be done" (102, 105). The relationship between TMs and musicians is based, in part, on actual activities of "looking after and looking out" rather than solely on predetermined tasks (Alacovska and Bissonnette 2019: 138). Such descriptions map neatly onto the artist-specific and artist-centric nature of the TM's role, and more generally to their positions as support personnel.

A key component of the TM-musician working relationship is that the people being looked after are adults that arguably would otherwise not require such oversight in a context outside of the particularities of live music and the status accorded artists. The notion that some people "need" such care speaks to a "difference in the relative value of different peoples' needs" (Tronto 1993: 116), which is evident in the working practices of touring. In turn, subordination, or the secondary status of support personnel, is maintained and reproduced in everyday life on the basis of who is cared for and who does the caring. This notion is further observable in the working lives of TMs, and road crews more generally, as my findings have shown, in how caregivers will "find that their needs to care for themselves come in conflict with the care that they must give to others" (109). In this way, people who have their needs cared-for by others are privileged, and privilege is determined, in part, by caring (116). The ability for musicians to have their needs attended to by others symbolises and reproduces the privileged status accorded them. Being a babysitter highlights the presence of care in the working relationship, and positions the TM in a caring role, however ambiguous or undesired it may be.

In the context of touring and the working relationship between TMs and musicians, the character of "babysitting" is difficult to precisely establish. The "lines for a babysitter are very blurred and I misunderstood them for a long time. I'm not saying I entirely understand them" (Joe 2019). Amy (2019) defined it as a "catch-all term in attempting to describe the part of our job that is not strictly day to day business and logistical activities, but more the 'personal' taking care of the people on tour end of things." In other words, those personal aspects of the job that TMs try to define and limit. Kim Hawes (2019: 290–291) stated that whenever she was on an airplane with a band, it was her usual practice to sit behind them to "make sure they were alright, that they didn't get in trouble with the airline staff, that they got the meals they had requested, and so forth." The term can also suggest how "difficult" an artist is and how much "care" or "attention" they need, as well as invoking the types of activities that require such attentiveness.

Some use the term to refer to the general practice of looking after musicians, inclusive of all aspects of the job, while others' use implies that babysitting refers to the need for a greater amount of effort or involvement on the part of TMs when attending to artists. Being a "babysitter" and "babysitting" can be understood as having a referential function that allows TMs to know their field, make sense of their working experiences and daily lives and understand their respective places (Conor 2013: 54). In "small-scale but important ways [they] reflect on the multiple functions and meanings of their work" (53). Such terms indicate that particular work activities are the "norm … and always have been" (44), but the manner in which they are used can also function as a critique of such practices. Rhetorical devices apply to the notion of the babysitter specifically and the experience and status of support personnel more generally because they are "used to understand … particular marginality" (53). While work associated with care, and those providing it, is often devalued (Tronto 1993: 114), the use of such terminology can function as a means to cope or resist. These terms create collegiality and shared purpose by encouraging professionalism while confronting marginalisation (Conor 2013: 53–54).

TMs use "babysitter" in several different contexts. The term functions as an expression of humour between TMs, and therefore as a marker of insider-outsider status. It is self-deprecating and facetious and used to refer to and make fun of key aspects of their job. The term is symbolic of an exclusive understanding and also functions as a coping mechanism and form of solidarity. Amy (2019) elaborated on how the term is used in this context:

> Tongue-in-cheek and used humorously between TMs. We joke all the time with each other and other people we work closely with on the touring side, who understand our jobs, about being 'glorified babysitters' but probably wouldn't use it in that particular context with someone who wasn't another touring person, or who we thought might take it as a complaint about our job.

Her caution about who she uses the term with, out of concern that it could be misunderstood or interpreted as a complaint, suggests the potential for it to be viewed as a criticism of the nature of the TM's job and of the artists with whom they work. This highlights that they use the term to invoke shared understanding within the safety of insider boundaries but are sensitive to how it can be taken out of context and reflect negatively on them. It also shows the importance of maintaining a reputation and the significance of the TM's working relationship with musicians in doing so.

Amy also stated that the term operates as a translatable reference point to explain the TM's role to people outside of the touring world[3]:

> For example, I've described my job as a TM before as 'part project manager, part travel agent, part accountant, part babysitter (or parent)' or something along those lines, because I can't think of a better way to encompass looking after the more personal aspects of the traveling party on a tour.

While Amy (2019) viewed the term as a suitable descriptor, Joe (2019) also spoke to the convenient use of the term in this context but emphasised the repercussions of its use. "When I'm trying to describe my job to somebody that doesn't understand it at all, I might use it but I would qualify it quite quickly, cause it patronises both them and me, it belittles what I do." As Amy did in reference to the term as an expression of humour, Joe also expressed caution when using the term, but for its potential to be offensive to both parties, it addresses rather than perceived as a complaint about the role. The term also circulates as a means to define the parameters of a TM's job. It is here that various interpretations of the term can be observed. Joe viewed the amount of babysitting that a particular artist requires as a factor in a decision to take a job. "When I take on a job I ask, you know how much babysitting ... do I have to do and get a sense of it before agreeing to take it. You know 'what's this artist like'." Consistent with the personal responsibilities associated with the job, Andy (2017) positioned babysitting as an element of the TM's role that needs to be clearly defined, and saw it in black and white terms.

> ... I think as a tour manager then you'd be looking to the ... artist management to say you know, okay, I'm not a babysitter, and that's the expectation, are you a babysitter or aren't you. On a different subject, I think when tour managers are being hired they need to define those roles, they need to define 'I am not a babysitter, I am managing the accounts, I'll get the band where they need to be, I'll take care of the promo, I'll sell merch.'

Amy (2019) recounted that it is used to clarify the nature of the tasks involved when working for a particular artist. She also described a clear separation of duties, but the lines of what constitutes babysitting differ from Andy's perception.

> ... if, for example, a management company had a TM job they were looking to hire for – if the job didn't involve any kind of advance coordination, or require me to do any tour accounting or technical work, but just look after the band and make sure they got from place to place per a schedule and making sure their day to day needs were covered, the person hiring would most likely describe the job along the lines of 'all the advance work has been done, you won't be settling shows, etc – you'd just be babysitting the band.'

The activities Amy described here, and outlined in the previous section, are among the most significant aspects of the TM's role. Andy sees getting a band to where they need to be as separate from babysitting whereas Amy's description suggests otherwise. A TM that was observed during fieldwork also referred to the communicative aspects of the job as babysitting when he described how he tells musicians to "be here at this time, no you can't do that, yes you can do this, be in the lobby at this time." TMs have contrasting understandings of the same term but collectively utilise it to convey different points. The term functions as a convenient, if problematic, shorthand, a comment on and means to define the nature of the work, and as self-deprecating humour, which is consistent with the broader workplace culture of touring. In this way, the term's functional usage is underpinned by how it can also "embody the ... frustrations and disappointments" that they encounter (Conor 2013: 53). It also shows that, despite the association of care with the TM's role, doing "caring work" does not require a "caring disposition." Rather, the act of care can be thought of "only in terms of a job" (Tronto 1993: 105).

The facetious use of the term suggests that TMs may find attending to such types of tasks unsatisfying. Joe (2019) explained that he prefers "working for somebody who is very definite about what they want, what they want to achieve, knows how to take care of themselves, knows how to take care of the business and sees me as a partner in making that happen." In other words, a position in which both parties are on more equal ground than the position of "babysitter" would imply or afford. He described how this is realised in his experience with the artist he currently works for:

> ... working with an artist who certainly doesn't need a babysitter. She needs somebody to enable all the things she wants to do and she needs constant attention to what she needs and what she needs to achieve. But she has very very strong ideas about herself and her career and her life and how she wants to live her life and how she wants to spend her money. Whether that's on her free time or on what crew she wants to spend it on or how she wants to tour or what hotels it is, it's her cash, it's her business and I manage it for her.
>
> *(Joe 2019)*

The TM is not there to "babysit" but to function as an assistant to help the artist realise goals.

Musicians may need and want the attention and involvement of the TM in the enactment and management of their live music careers. Such business-oriented and professional concerns benefit from the assistance of a knowledgeable and experienced TM and are markedly different than the kinds of activities that babysitting implies.

TMs are careful not to use the term in front of musicians, not even in the context of humour. This shows again that they make efforts to protect musicians. Artists could "take that really badly" (Joe 2019) and "it's one of those things that maybe an artist might take personally, or think you're being disparaging about them" (Amy 2019). Artists also do not refer to TMs as babysitters and it is unclear the extent to which they are even aware of the term.

The issue of the artist's perspective represents a reversal to that of musicians seeing TMs as parental figures. Musicians will call TMs their "tour mom" or liken them to a "parent," particularly those with whom they have worked for a long time and come to trust. For example, members of the band Queen referred to TM Gerry Stickells, following his death, as a "father figure, great friend and teacher, and an island of calm in the midst of chaos" (Sandomir 2019). Guy Garvey, of the band Elbow, positioned the band's TM, Tom Piper, in a parental role by describing him as "dad to about 15 people. Anything we don't know, we go to him" (Pattenden 2002: 84). Amy (2019) described it as a "term of endearment" and as a signal that the TM is doing well and is trusted. A feature in *Q* magazine supported this statement, articulating that the "best" TMs "become father figures to their bands" (Pattenden 2002: 84). Moving from the temporary status of babysitter to one of parent is an interesting discursive switch that indicates change and continuity in the nature of the relationship. It also represents further similarity between TMs and tour guides, who have been referred to as "surrogate parents" (Reisinger and Steiner 2006: 482).

Such language is also evident in other sectors of the music industries. PR staff use similar language to describe responsibilities that involve being at musicians' "beck and call" and as having to "'nanny people' … it was just like running a bloody kindergarten, except that the people you were dealing with weren't as nice as kids" (Steward and Garratt 1984: 68, Negus [1992] 2011: 117).

The term babysitter has deeper associations and connotations. It invokes gendered perspectives on work and correlates with ideas of youth, popular music and musicians' identities. Its use amongst TMs, and their perceptions of and responses to it, refer to, reproduce and problematise these associations. Though tour management is male-dominated, aspects of the job involve tasks that are traditionally gendered feminine (see also Kielich 2021). Care is often thought of as the domain of women and is associated with service and people occupying marginalised or secondary status (Tronto 1993: 112–113). The gender imbalance in live music, and its continuity, can distance or obscure the feminine aspects or associations of work activities. The term babysitter makes those elements explicit.

Babysitting is a job that both young girls and boys typically do; however, it is primarily considered a form of feminised labour given its emphasis on childcare (see Forman-Brunell 2009). In this way, the term refers to feminised labour and the manner in which TMs engage with it provides a comment on and insights into how gender functions on tour and in their occupational identities. Women and men research respondents had contrasting attitudes and perspectives and expressed different preferences regarding usage of the term. The former utilised the term facetiously as a form of workplace humour, while expressing concern over the term being misread as a complaint about TMs' line of work and was sensitive to how it could offend artists. The notion of babysitting was taken for granted as a normal and expected part of the role. The latter did not take such activities for granted nor accept them as an inevitable part of a TM's job. They were critical of the term, described it as "patronizing," expressed discomfort with it and a dislike of that role. However, they admitted to using it as an "easy handle" despite not liking it. Such views, along with the manner in which male TMs described negotiating the place of babysitting in a given job, can be seen as a potential resistance to performing or being associated with feminised labour. It is another means of creating and maintaining "distance from the feminine" (Pullen and Simpson 2009: 564). Scholars have argued that male gender categories tend to be situated in "a work-based (rather than family-based) culture of manhood" (Segal 1990: 94, Connell 1995: 93, Den Tandt 2005: 380). Related to this, one element of "work-based masculinity" depends on developing "[t]echnical knowledge and expertise" (Segal 1990: 96, Den Tandt 2005: 380). While TMs do accumulate various forms of knowledge and expertise, the expectations associated with "work-based masculinity" differ from their skillset. In this way, male TMs may seek alternate ways to affirm masculinity in the workplace. The very identification of the term "babysitter" as a form of humour also functions this way, as a means to reappropriate its use and mask its meaning, similar to the wider use of humour in the workplace culture of touring.

However, the connection between babysitting and care also reveals masculine associations with the TM's role. The notion of "taking care of," which is a key component in their daily working lives with musicians, is more strongly linked to men and masculinity (Tronto 1993: 115). TMs do not so much focus on the needs of musicians – as in "caring about" – but rather act to address them (106) in the context of their roles. Through daily activities on tour, TMs "take care of" musicians in the sense that they assume "some responsibility for the identified need and determin[e] how to respond to it" (ibid.). In this sense, care is a practice and activity that involves agency and obligation

(106). However, their proximity and direct involvement in the everyday lives of musicians realigns them with forms of feminised labour. To illustrate how this operates in the working lives of TMs, doctors, a typically male-dominated position, "take care of" patients and gain prestige from their status. Following this, TMs are at the top of the touring party's hierarchy, are knowledgeable and experienced, make the plans, give directions and solve problems when necessary – and can maintain detachment by resisting a caring disposition. In contrast to doctors, nurses, traditionally gendered feminine, or orderlies and lab technicians, have the duty of hands-on care (115). TMs also engage in a type of "care-giving" through actions that satisfy needs and involve physical work and direct contact (107). TMs simultaneously "take care of" and are tasked with attending to the everyday activities, or the actual "care." Following this, care is often associated with feminised labour but can also be regarded as an emotionally detached form of work. In this way, the "caring" activities of TMs highlight that tensions exist with notions of care work and suggests that such activities are shaped and defined according to the people and roles that enact such responsibilities.

The term "babysitter" can also be understood within the broader context of rock and popular music. Its usage is significant in view of musicians and their identities and can be seen as an extension, reflection and reproduction of the connection between youth and popular music. In the specific context of rock music culture, "growing up" has been viewed as a life stage that is best avoided. Rock represented an "expressive context for strategies of youth identity" where "growing up" was only desirable as a means to access sex, drugs and rock and roll, but was deemed undesirable because becoming an adult risks becoming boring (Grossberg 1987). Musicians have often cited this as a motivation to form a rock band. In the Rolling Stones documentary *Crossfire Hurricane* (Morgen 2012), Mick Jagger explained that getting into a successful band was about "not growing up." This notion can be further observed in media coverage that summarises musicians' autobiographies as "Boys' Own Tales: Four Rock Stars Who Refused to Grow Up" (Maconie 2014). In addition, bad or youthful behaviour became subsequently associated and accepted. In other words, it is desirable to grow up just enough to access particular interests, and then the status of musician functions to preserve the youthful identity and permission to behave accordingly. As Leonard (2017 [2007]: 62) notes, these notions become integrated and normalised in rock culture and associated industry norms. She argues that "the concept of artistic temperament and behaviour patterns" – such as a perceived refusal to grow up and the tolerance of bad behaviour – and the "split between creativity and commerce, form part of a rock discourse that pervades the industry and affects business practice." This chapter has demonstrated some of the ways in which this operates in live music. One can wonder, however, if being free and young are concerns specific to men, and if male musicians in particular are invested in the idea of not growing up.

As a category, rock stars are "to a greater or lesser extent, exempt from the rules of adulthood in that they are expected to be rebellious and badly-behaved" (Strong 2016: 129). In the context of touring, such exemption is not strictly related to "bad" behaviour but can be seen in more generalised terms around what is expected and accepted. As such, it follows that related behaviours would be tolerated and accommodated by personnel around musicians, in the same way that artists are granted special privileges based on their status (Becker 1982). In this way, TMs unintentionally reproduce these notions through participation in some of the conventions of touring and the usage of the term "babysitter." Furthermore, the potential for musicians to be infantilised by the conventions of touring interacts with these ideas, which also create conditions for such effects to be taken for granted. At the same time, such behaviours are those that TMs often try to avoid having to attend to when negotiating and accepting a position. In this way, the term babysitter can be understood as a comment and critique. Its potential to have a negative connotation amongst TMs is another reminder of the insider-outsider boundaries of touring and road crew members. Such notions of youth, behaviour and musicians' identities are among the myths and clichés of rock music yet the realities of them are a normal part of everyday life for insiders. The following will provide a summary and conclusion of this chapter.

Conclusion

This chapter has explored and examined how TMs look after musicians on tour and has showed that doing so is an integral component in the realisation of live music events. Following on from live music being the purpose and objective of a tour, it has established that the artist is the most important focus of a TM's job. The role is both artist-specific and artistic-centric and TMs' working lives are shaped by the activities that accompany the needs and preferences of musicians. The chapter has demonstrated that the TM's daily activities are integral by highlighting that they protect, provide, support and care for musicians. The instrumental, social, interactional and communicative components of their roles illustrate how they do so. TMs' ability to alleviate tasks, enable focus and reduce stress for musicians is an essential factor in a pressurised and mobile workplace. These efforts are not without effects, and they can enable musicians to become dependent on the personnel around them, which is part of a larger pattern in the music industries. The TM's role is at once important to supporting musicians at the same time that it reinforces musicians' importance to live music. Yet their centrality to live events also means that the efforts of TMs benefit musicians while creating compromise in their own working lives. As such, the effects on the people responsible for such integral tasks are taken for granted. Daily life on tour is based on activities and conventions that reproduce the special status accorded musicians and the secondary status of TMs. At the same time, musicians and the live music industry ultimately depend on the efforts of TMs to realise events, which further substantiates that they are primary workers in supporting roles.

Following this, TMs have an unstable relationship with power and authority that is both a product of their positions as middle managers and based on the nature of their working relationships with musicians. The role of the TM ultimately moves between positions of authority and subordination. Their dual role as a service provider and as a leader can be at odds with musicians' position as their "boss" due to the ways in which this creates an inherent tension that is difficult to overcome. The working dynamic can, however, be successful if mutual trust and understanding exist between them. This chapter has established that the TM's role is adjustable, ambiguous and open to interpretation. As such, it is essential for TMs to define the parameters and set limits as to the types of activities it will entail. TMs use the term "babysitter" to joke about – and as shorthand for – the nature of their roles yet also express ambivalence or dislike of the term. It has the potential to undermine their positions and illuminates gendered aspects of their tasks that are otherwise masked by the male-dominated workforce. As such, the term in its use and resistance highlights important features that reflect and shape the working lives of TMs and their occupational identities.

In analysing the manner in which TMs look after musicians, this chapter has shown that such tasks are necessary to make a concert happen and that the manner in which they do so is a significant aspect of live music events.

Notes

1 Gross and Musgrave (2020: 129) provide the following definition: "In legal terms, the duty of care confers a legal responsibility to act in accordance with an ideal of reasonable care so as to prevent the occurrence of 'foreseeable' personal harm to others, which can include mental as well as physical harm."
2 See Cavicchi's (1998: 67) study of Bruce Springsteen fans for further evidence that a pre-paid return envelope is an effective way to acquire an autograph.
3 See also Douglas (2021: 238).

References

Alacovska, Ana and Joëlle Bissonnette. 2019. "Care-ful Work: An Ethics of Care Approach to Contingent Labour in the Creative Industries." *Journal of Business Ethics* 169: 135–151.

Anicich, Eric M. and Jacob B. Hirsh. 2017a. "The Psychology of Middle Power: Vertical Code-Switching, Role Conflict, and Behavioral Inhibition." *Academy of Management Review* 42(4): 659–682.

Anicich, Eric M. and Jacob B. Hirsh. 2017b. "Why Being a Middle Manager Is So Exhausting." *Harvard Business Review* (March 22): https://hbr.org/2017/03/why-being-a-middle-manager-is-so-exhausting.

Atkins, Martin. 2007. *Tour:Smart: And Break the Band*. 1st ed. Chicago, IL: Chicago Review Press.

Attali, Jacques. 1985. *Noise: The Political Economy of Music*. Minneapolis: University of Minnesota Press.

Auslander, Philip. 2008. *Liveness: Performance in a Mediatized Culture*. New York: Routledge.

Barr, Ken. 2009. *We Are the Road Crew, Vol. 1*. Scotts Valley, CA: CreateSpace Independent Publishing Platform.

Baym, Nancy K. 2018. *Playing to the Crowd: Musicians, Audiences, and the Intimate Work of Connection*. New York: New York University Press.

Bayton, Mavis 1998. *Frock Rock: Women Performing Popular Music*. Oxford: Oxford University Press.

Beaumont-Thomas, Ben. 2017. "Belle and Sebastian Accidentally Leave Drummer in Pyjamas in Walmart." *The Guardian* (17 August): https://www.theguardian.com/music/2017/aug/16/belle-and-sebastian-accidentally-leave-drummer-in-pyjamas-in-walmart.

Becker, Howard S. 1982. *Art Worlds*. Berkeley: University of California Press.

Bennett, H. Stith. (1980) 2017. *On Becoming a Rock Musician*. New York: Columbia University Press.

Black, Grant C., Mark A. Fox and Paul Kochanowski. 2007. "Concert Tour Success in North America: An Examination of the Top 100 Tours from 1997 to 2005." *Popular Music and Society* 30(2): 149–172.

Cavicchi, Daniel. 1998. *Tramps Like Us: Music and Meaning Among Springsteen Fans*. Oxford: Oxford University Press.

Cohen, Erik. 1985. "The Tourist Guide: The Origins, Structure and Dynamics of a Role." *Annals of Tourism Research* 12: 5–29.

Conor, Bridget. 2013. "Hired Hands, Liars, Schmucks: Histories of Screenwriting Work and Workers in Contemporary Screen Production." In *Theorizing Cultural Work: Labour, Continuity and Change in the Cultural and Creative Industries*, eds. Mark Banks, Rosalind Gill and Sarah Taylor, 44–55. London: Routledge.

Cook, Malcolm. 2011. *Cook's Tours: Tales of a Tour Manager*. York: Music Mentor Books.

Connell, R.W. 1995. *Masculinities*. Berkeley: University of California Press.

Curtin, Michael and Kevin Sanson, eds. 2017. *Voices of Labor: Creativity, Craft, and Conflict in Global Hollywood*. Oakland: University of California Press.

Den Tandt, Christophe. 2005. "Musical Craftsmanship, Gender, and Cultural Capital in Classic Rock." In *Reading Without Maps?: Cultural Landmarks in a Post-Canonical Age: A Tribute to Gilbert Debusscher*, eds. Marc Maufort and Christophe Den Tandt, 379–399. Brussels: P.I.E. Peter Lang.

Ehrlich, Cyril. 1985. *The Music Profession in Britain Since the Eighteenth Century: A Social History*. Oxford: Clarendon Press.

Fonarow, Wendy. 2006. *Empire of Dirt: The Aesthetics and Rituals of British Indie Music*. Middletown, CT: Wesleyan University Press.

Forman-Brunell, Miriam. 2009. *Babysitter: An American History*. New York: New York University Press.

Forsyth, Alasdair, Jemma Lennox, and Carol Emslie. 2016. "'That's Cool, You're a Musician and You Drink': Exploring Entertainers' Accounts of Their Unique Workplace Relationship with Alcohol." *International Journal of Drug Policy* 36(2): 85–94.

Frith, Simon, Matt Brennan, Martin Cloonan and Emma Webster. 2013. *The History of Live Music in Britain, Volume I: 1950–1967*. Farnham: Ashgate.

Frith, Simon, Matt Brennan, Martin Cloonan and Emma Webster. 2019. *The History of Live Music in Britain, Volume II: 1968–1984*. London: Routledge.

Gorman, Clem. 1978. *Backstage Rock: Behind the Scene with the Bands*. London: Pan Books.

Groce, Stephen B. 1991. "What's the Buzz?: Rethinking the Meanings and Uses of Alcohol and Other Drugs Among Small-Time Rock 'n' Roll Musicians." *Deviant Behavior* 12(4): 361–384.

Gross, Sally Anne and George Musgrave. 2020. *Can Music Make You Sick? Measuring the Price of Musical Ambition*. London: University of Westminster Press.

Grossberg, Lawrence. 1987. "Rock and Roll in Search of an Audience." In *Popular Music and Communication*, ed. James Lull, 175–197. Newbury Park, CA: Sage.

Hawes, Kim. 2019. *Confessions of a Female Tour Manager*. Independently Published.

Hawes, Kim. 2023. *Lipstick and Leather: On the Road with the World's Most Notorious Rock Stars*. Muir of Ord, Scotland: Sandstone Press, Ltd.

Hart, Colin. 2011. *A Hart Life*. Bedford: Wymer Publishing.

Held, Virginia. 2006. *The Ethics of Care: Personal, Political, and Global*. Oxford: Oxford University Press.

Hesmondhalgh, David. 2013. *The Cultural Industries*. 3rd ed. London: Sage.

Hesmondhalgh, David and Sarah Baker. 2011. *Creative Labour: Media Work in Three Cultural Industries*. New York: Routledge.

Holloway, J. Christopher. 1981 "The Guided Tour: A Sociological Approach." *Annals of Tourism Research* 8(3): 377–402.

Hracs, Brian J. 2015. "Working Harder and Working Smarter: The Survival Strategies of Contemporary Independent Musicians." In *The Production and Consumption of Music in the Digital Age*, eds. Brian J. Hracs, Michael Seman and Tarek E. Virani, 48–62. London: Routledge.

Jones, Michael. 2012. *The Music Industries: From Conception to Consumption*. Basingstoke: Palgrave MacMillan.

Kankkunen, Tina Forsberg. 2014. "Access to Networks in Genderized Contexts: The Construction of Hierarchical Networks and Inequalities in Feminized, Caring and Masculinized, Technical Occupations." *Gender, Work & Organization* 21(4): 340–352.

Kearney, Mary Celeste. 2017. *Gender and Rock*. Oxford: Oxford University Press.

Kielich, Gabrielle. 2015. "The 'Resort' Studio: An Introduction to the History and Culture of the Residential Recording Studio." MA Thesis, Carleton University.

Kielich, Gabrielle. 2021. "Fulfilling the Hospitality Rider: Working Practices and Issues in a Tour's Supply Chain." In *Researching Live Music: Gigs, Tours, Concerts and Festivals*, eds. Chris Anderton and Sergio Pisfil, 115–126. London: Taylor & Francis/Routledge.

Maconie, Stuart. 2014. "Boys' Own Tales: Four Rock Stars Who Refused to Grow Up." *New Statesman* (19 December): https://www.newstatesman.com/culture/2014/12/boys-own-tales-four-rock-stars-who-refused-grow.

McKinna, Daniel R. 2014. "The Touring Musician: Repetition and Authenticity in Performance." *IASPM@Journal* 4(1): 56–72.

Morgen, Brett, dir. 2012. *Crossfire Hurricane*. Home Box Office (HBO). Television.

Negus, Keith. (1992) 2011. *Producing Pop: Culture and Conflict in the Popular Music Industry*. Mountain View, CA: Creative Commons.

Nonaka, Ikujiro. 1994. "A Dynamic Theory of Knowledge Creation." *Organization Science* 5: 14–37.

Pattenden, Mike. 2002. "Is This the Worst Job in the World?" *Q* (March): 80–84.

Pullen, Alison and Ruth Simpson. 2009. "Managing Difference in Feminized Work: Men, Otherness and Social Practice." *Human Relations* 62(4): 561–587.

Raeburn, Susan D. 1987. "Occupational Stress and Coping in a Sample of Professional Rock Musicians, Part 1." *Medical Problems of Performing Artists* 2: 41–48.

Raven, Ben. 2015. "3 Tour Managers Tell Us the Real Side of a Life On the Road." *Mixmag* (13 April): http://mixmag.net/feature/dream-team.

Reisinger, Yvette and Carol Steiner. 2006. "Reconceptualising Interpretation: The Role of Tour Guides in Authentic Tourism." *Current Issues in Tourism* 9(6): 481–498.

Reynolds, Andy. 2008. *The Tour Book: How to Get Your Music on the Road*. Boston, MA: Cengage Learning.

Reynolds, Andy. 2012. *Roadie, Inc.: How to Gain and Keep a Career in the Live Music Business*. 2nd ed. Scotts Valley, CA: CreateSpace Independent Publishing Platform.

Sandomir, Richard. 2019. "Gerry Stickells, Who Helped Make Rock Shows Big, Dies at 76." *New York Times* (3 April): https://www.nytimes.com/2019/04/03/obituaries/gerry-stickells-dead.html.

Schmidt, Catherine J. 1979. "The Guided Tour: Insulated Adventure." *Urban Life* 7(4): 441–467.

Shuker, Roy. 2008. *Understanding Popular Music Culture*. 3rd ed. London: Routledge.

Simpson, Dave. 2010. "Rock 'n' Roll Jobs Explained." *The Guardian* (15 April): https://www.theguardian.com/music/2010/apr/15/rock-n-roll-jobs-explained.

Simpson, Dave. 2015. "How to Put on a Mega-Gig: The Tour Manager's Story." *The Guardian* (14 August): https://www.theguardian.com/music/musicblog/2015/aug/14/how-to-put-on-a-mega-gig-the-tour-managers-story.

Small, Christopher. 1999. *Musicking: The Meanings of Performing and Listening*. Hanover: Wesleyan University Press.

Stahl, Matt. 2013. *Unfree Masters: Recording Artists and the Politics of Work*. Durham, NC: Duke University Press.

Steward, Sue and Sheryl Garratt. 1984. *Signed, Sealed, and Delivered: True Life Stories of Women in Pop*. London: Pluto Press.

Stokes, Geoffrey. 1976. *Star-Making Machinery: Inside the Business of Rock and Roll*. New York: Vintage Books.

Strong, Catherine. 2016. "'I'd Stage-Dive, But I'm Far Too Elderly': Courtney Love and Expectations of Femininity and Ageing." In *"Rock On": Women, Ageing and Popular Music*, eds. Ros Jennings and Abigail Gardner, 124–137. London: Routledge.

Tronto, Joan C. 1993. *Moral Boundaries: A Political Argument for an Ethic of Care*. New York: Routledge.

Tuchman, Gaye. 1973. "Making News by Doing Work: Routinizing the Unexpected." *American Journal of Sociology* 79(1): 110–131.

Webster, Emma. 2011. "Promoting Live Music in the UK: A Behind-the-Scenes Ethnography." PhD diss., University of Glasgow.

Williams, Raymond. 1983. *Keywords: A Vocabulary of Culture and Society*. Revised ed. New York: Oxford University Press.

Williamson, John and Martin Cloonan. 2016. *Players' Work Time: A History of the British Musicians' Union, 1893–2013*. Manchester: Manchester University Press.

Wooldridge, Bill, Torsten Schmid and Steven W. Floyd. 2008. "The Middle Management Perspective on Strategy Process: Contributions, Synthesis, and Future Research." *Journal of Management* 34(6): 1190–1221.

Workman, Mark. 2012. *One for the Road: How to Be a Tour Manager*. Road Crew Books.

Wright, James "Tappy" and Rod Weinberg. 2009. *Rock Roadie: Backstage and Confidential with Hendrix, Elvis, The Animals, Tina Turner and an All-Star Cast*. London: JR Books.

Interviews

Amy. Tour Manager. Skype Interview, 9 July 2018. Email Correspondence, 18 June 2019.

Andy. Tour Manager and Sound Engineer. In-Person Interview, 18 April 2017.

Joe. Tour and Production Manager. In-Person Interview, 2 May 2017, 14 May 2018. Skype Interview, 28 August 2019.

CONCLUSION

This book has demonstrated the ways in which the road crew is a key group of live music's support personnel. It has examined their integral roles in the realisation of concert events and explored the variety of factors that shape their working lives and occupational identities. Research findings presented here have established that road crews are groups of specialised workers, each with important individual roles, who collectively participate in a shared purpose. I have also put forth evidence that renders the term "roadie" a problematic descriptor and inaccurate reflection of what road crew members do. This book has provided insights into the ways they learn their roles, gain access to the live music industry and maintain work. In particular, it has highlighted that their career paths are carved through informal processes and practices and greatly assisted by a network of contacts.

This study has placed particular emphasis on the significance of road crew members as support personnel, and how such a status shapes their working lives. Due to their active and essential roles in live music, I have argued that crew members are primary workers in supporting roles and destabilise notions of hierarchy in the music industries. Research findings have established that road crews are marked by an underrepresentation of women and minimal diversity, consistent with the wider music industries. Women encounter a range of challenges related to access to training and treatment in the workplace, indicating some of the ways in which gender inequality functions in live music.

In examining road crews, this book has also highlighted the importance of touring to the realisation of live music and its significance in their working lives. It has demonstrated that crew members are similar to many types of workers in the cultural industries in that their working lives are characterised

DOI: 10.4324/9781003303046-7

by a long-hours culture. Road crew members are part of the complex mobile world of touring, which is shaped and maintained by norms, expectations and a particular workplace culture. They learn to manage working and living together, form bonds and share humour to ensure the smooth continuity of life on the road. Touring is permeated by a highly masculinist culture, which is connected to the maintenance of the male domination of road crews, the gender inequality in live music and the related difficulties that women encounter, and to elements of the norms and expectations around personal conduct. Crew members also face challenging effects of itinerant work in the form of disorientation and difficulties in the maintenance of mental and physical health, both on and off the road.

This book has shown and analysed the significance of tour mangers' roles in the realisation of live music, and in particular relation to their working lives with musicians. TMs play an integral role in looking after musicians and creating the space that enables them to prepare for performance in a pressurised workplace. Such activities can result in dependencies and the notion that TMs are "babysitters." They also show that care is a feature of the job responsibilities of TMs, based on how creative work ceases to be individualised and through the foregrounding of musicians' needs. I have argued that the notion of care, and the types of caring work that feature in tour, advances the understanding of the role and occupational identities of support personnel. In particular, care work and support work are associated with the potential to be subordinated or viewed as secondary, yet also to be actively involved in taking care of individuals and necessary tasks, and both require prioritising the needs of others, which is determined by and reproduces status and privilege. In these ways, road crews and the norms and practices of touring are significant source of insight for further understanding support personnel and the way care functions in the workplace.

A major aim of this book has been to provide groundwork about a group of support personnel and the characteristics of their working lives that had previously been minimally explored. In doing so, this book makes a contribution to live music studies by widening the scope of understanding about road crews and touring and in its consideration and analysis of a broader set of people and activities that are integral to the realisation of concert events. In particular, I put forth three major points about live music and its study in this conclusion.

First, we need to see the ways in which live music events, particularly those that "fall[] [within] a wider temporal series of events … called [a] tour" (Weinstein [1991] 2000: 203), are brought into being through work that is both well-defined and negotiated on a daily basis. Live music events are realised, to some extent, through legal contracts and formal agreements at the same time that they are the result of informal, off-the-record discussions and disagreements. They are brought about through careful planning and last-minute decisions. Clearly established conventions and expectations matter as much to understanding the live music sector as do sudden, unanticipated circumstances and the manner in which they are handled. Particular work activities may be presented one way on paper, but in practice be enacted according to very different criteria. The significance of interpersonal relations in the live music industry renders daily negotiations between workers a central site of analysis in understanding the labour involved in concert events. In addition, ambiguous and contextual job descriptions, the place of artist's preferences and changing personnel within crews and at the local level mean there is much to be learned from factors that are difficult to systematise.

Second, live music is so often understood as involving special, magical moments (Black, Fox and Kochanowski 2007: 154–155, Frith 2007: 14, Brown and Knox 2017: 238–241), but is, in fact, realised through the most repetitive and mundane of activities. The mobile nature of touring means that venues function as hubs to sustain basic needs and daily life. Successfully putting a show together depends on access to a supply chain (see also Kielich 2021), the circulation of documents and effective communication. The smooth continuity of a tour places importance on mutual respect and tolerance given the close proximity in which personnel live and work as they move from one show day to the next. Being on tour is also enhanced by maintaining morale, a repetitive act in and of itself given the ubiquitous use of humour within the workplace culture. In these ways, the unspectacular activities that constitute live events are rich sources of understanding. While some may argue that awareness of such tasks diminishes the effect of the spectacle (see White 2014: 1), it can also be said that looking behind the curtain ultimately makes the result more impressive.

Third, tours are particular ways of moving through the world by groups of people brought together in sets of relationships that are worked out along the way. We must recognise the activities that occur both within and between venues as important sources of analysis, as well as the role of mobility in shaping the lives, experiences and identities of workers on tour (see also Nóvoa 2012). Understanding live music means making sense of its workplace culture, and the norms, expectations and ambiguities that touring personnel encounter. It means locating the power dynamics, gender imbalances and interpersonal relations that are significant to the everyday conditions of working on tour, and how personnel make sense of and address them. The efforts and experiences of workers on tour are part of live music events and are influenced by the process of touring.

In its goal of providing groundwork, this study could not be comprehensive nor address all aspects related to road crews and touring. It is my hope that others will find road crews and touring to be inspiring and stimulating objects of research as there is much room to further develop these topics. This study was a contemporary one, but the past, present and the immediate future are all worthy of a more comprehensive study. A longer historical account of road crews – or "roadies" – and how their working lives have changed or stayed the same, along with the terminology used to describe them, would extend the scope of understanding and create insightful connections. Studies that explore diversity amongst workers on road crews, particularly in relation to gender diversity and nonconformity as well as sexual orientation, would contribute to an important dialogue and be of widespread relevance. Similarly, research is needed that extends beyond the parameters of Western and English-speaking crew members and tours.

In-depth studies of specific roles on road crews, such as production managers, guitar technicians or riggers, would enhance the understanding of support personnel broadly, and their individual character and significance specifically, in the music and cultural industries. Such research would contribute further understanding of the individual and collective natures of road crews in the realisation of live music events. Focused studies of the work activities of other types of crew members would also widen the scope and demonstrate more clearly how the issue of care functions and factors into the culture of touring and the working relationships with musicians. Differences and similarities, along with patterns across roles, could be located. Additionally, a study that explores the working relationships between road crews and musicians from the perspective of the latter, as well as their daily experiences on tour, would offer useful comparison and insights. Furthermore, studies that identify and explain the onstage communication techniques between crew members and musicians, which appear to be highly subjective and informal, would contribute a deeper understanding of the working practices of live music as well as the factors that make a concert happen.

In closing, road crew members may have commonly been described as and known for being the "unsung heroes" of live music. The research and analysis presented in this book has hopefully generated new perspectives on this group of touring personnel. Perhaps we can now begin to shift our thinking away from what is not known or appreciated about them towards an understanding and recognition of the significant, essential and meaningful contributions they make to the realisation of concert events on a daily basis.

References

Black, Grant C., Mark A. Fox and Paul Kochanowski. 2007. "Concert Tour Success in North America: An Examination of the Top 100 Tours from 1997 to 2005." *Popular Music and Society* 30(2): 149–172.

Brown, Steven Caldwell and Don Knox. 2017. "Why Go to Pop Concerts? The Motivations Behind Live Music Attendance." *Musicae Scientiae* 21(3): 233–249.

Frith, Simon. 2007. "Live Music Matters." *Scottish Music Review* 1(1): 1–17.
Kielich, Gabrielle. 2021. "Fulfilling the Hospitality Rider: Working Practices and Issues in a Tour's Supply Chain." In *Researching Live Music: Gigs, Tours, Concerts and Festivals*, eds. Chris Anderton and Sergio Pisfil, 115–126. London: Taylor & Francis/Routledge.
Nóvoa, André. 2012. "Musicians on the Move: Mobilities and Identities of a Band on the Road." *Mobilities* 7(3): 349–368.
Weinstein, Deena. (1991) 2000. *Heavy Metal: The Music and Its Culture*. Cambridge, MA: DaCapo Press.
White, Timothy R. 2014. *Blue Collar Broadway: The Craft and Industry of American Theater*. Philadelphia: University of Pennsylvania Press.

INDEX

Note: *Italic* page numbers refer to boxes.

Printed in the United States
by Baker & Taylor Publisher Services